SENSOR FOUNDATIONS SERIES

Accelerometers

By Jay Esfandyari, Ph.D.

Published by CKing Sensors Press
An imprint of CKing Sensors LLC

Paperback ISBN: 979-8-9957174-0-9

This book provides a beginner-friendly introduction to accelerometers. It does not replace manufacturer datasheets, application notes, or professional engineering guidance.

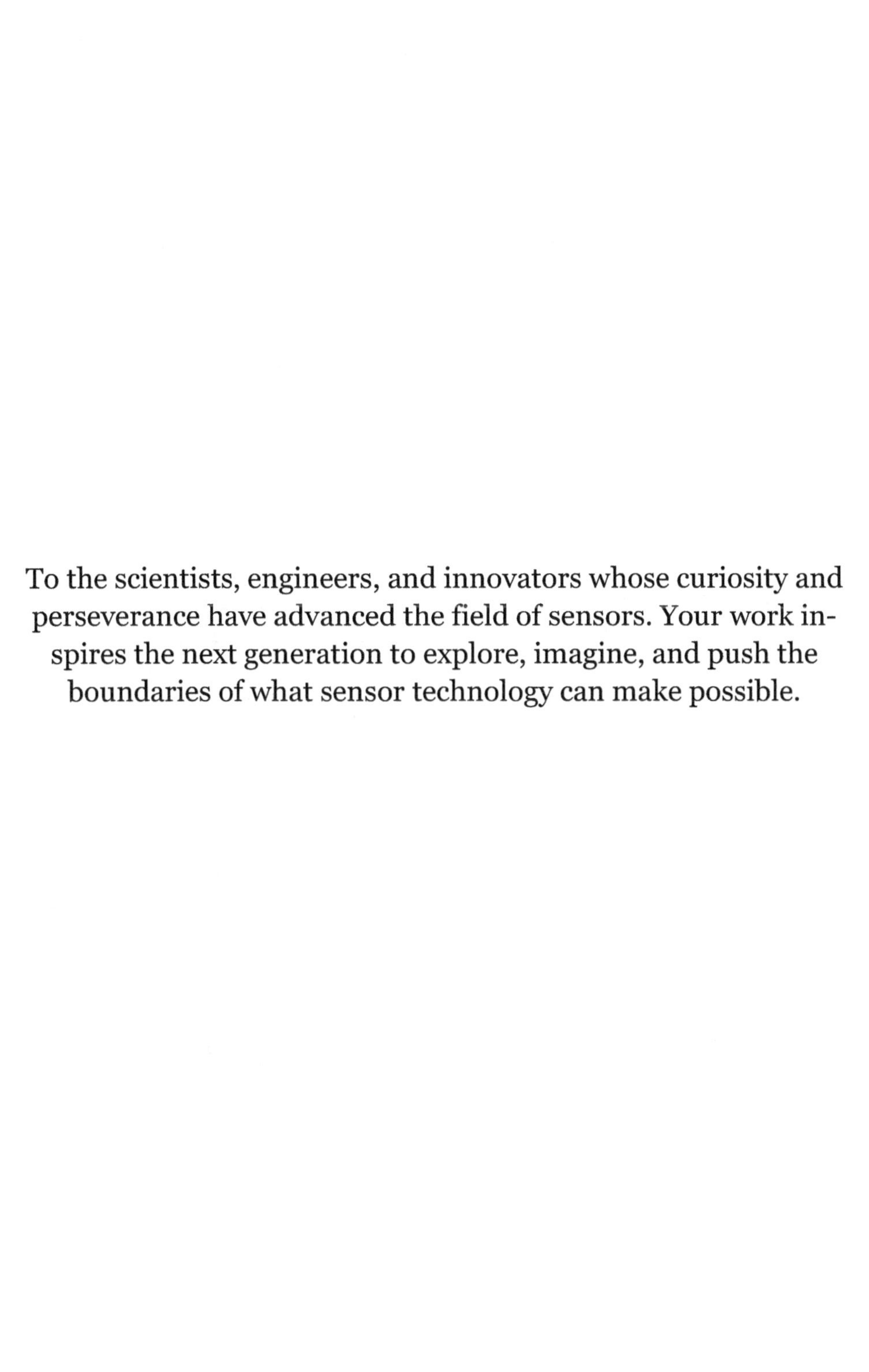

To the scientists, engineers, and innovators whose curiosity and perseverance have advanced the field of sensors. Your work inspires the next generation to explore, imagine, and push the boundaries of what sensor technology can make possible.

Introduction 1
CHAPTER 1 Accelerometer Fundamentals 4
1.1 What Does It Mean to Measure Acceleration 4
1.2 Types of Accelerometers 8
1.2.1 Mechanical (Mass-Spring) Accelerometers 8
1.2.2 Capacitive Accelerometers (Most Common in MEMS) 9
1.2.3 Piezoelectric Accelerometers 10
1.2.4 Piezoresistive Accelerometers 10
1.2.5 Thermal (Convection) Accelerometers 11
1.2.6 Servo (Force-Balance) Accelerometers 11
1.2.7 Resonant Accelerometers 12
1.2.8 Optical Accelerometers 12
1.2.9 Emerging Accelerometer Technologies 12
1.3 MEMS Accelerometers 13
1.3.1 Miniaturization 14
1.3.2 High Volume Manufacturing 15
1.3.3 Compatibility With Electronic Integration 16
1.3.4 Low Power Consumption 16
1.3.5 High Reliability and Robustness 16
1.3.6 Why the Industry Chose Capacitive MEMS 17
1.4 Capacitive MEMS Accelerometers 17
1.4.1 The Principle of Variable Capacitance 17
1.4.2 Why MEMS Makes Capacitive Sensing Possible 20
1.5 A Simple Mathematical Intuition Behind Acceleration 21
1.5.1 Acceleration as the Change of Velocity Over Time 21
1.5.2 Gravity as a Constant Acceleration 22
1.5.3 Relating Acceleration to Proof Mass Displacement .. 23
1.5.4 Relating Tilt to Acceleration 24

1.5.5 Vibrations and Frequency Response 25
1.6 Real-World Applications of Accelerometers.................... 26
1.7 Takeaways 28
CHAPTER 2 Inside a MEMS Accelerometer 30
2.1 Core Principle: From Mechanical Motion to Electrical Signal 31
2.1.1 From Mechanical Motion to Electrical Signals and Digital Data 31
2.1.2 Why Mechanical Motion Is So Small, Yet So Useful 32
2.1.3 Why Orientation Is Measurable Even Without Motion 32
2.2 Inside a Capacitive MEMS Accelerometer...................... 32
2.2.1 The Proof Mass: A Tiny Inertial Element................ 33
2.2.2 Suspension Springs: Allowing Motion and Setting Sensitivity 34
2.2.3 Comb-Finger Structures: Capacitive Sensing..........36
2.2.4 Differential Capacitors: The Engine of High Precision 38
2.2.5 Air Damping Around the Proof Mass...................... 39
2.2.6 Mechanical Stops: Protecting the Device From Shock 39
2.2.7 Multi-Axis Sensing: Measuring X, Y, and Z in a Single Package 40
2.3 The Signal Chain: Capacitance to Digital Output 41
2.3.1 The Capacitive Domain: Sensing the Change.......... 42
2.3.2 Charge-Based Measurement: Converting ΔC Into Voltage 42
2.3.3 The Analog Front-End (AFE): Amplification and Filtering 43
2.3.4 The ADC: Turning Analog Signals Into Numbers ... 44

2.3.5 Digital Signal Processing and Self-Test 44
2.3.6 Output Formatting and Data Delivery 45
2.4 Internal Dynamics That Shape Performance 45
2.4.1 Mechanical Damping: Controlling How the Proof Mass Moves 46
2.4.2 Noise: Why Accelerometers Can Never Be Perfectly Quiet 46
2.4.3 Resonance and Natural Frequency in MEMS Structures 48
2.4.4 Linearity: A Crucial Feature for Accurate Measurements 49
2.4.5 Cross-Axis Behavior: Minimizing Interference........50
2.4.6 Bandwidth and Settling Behavior 51
2.4.7 Stability Over Time: Drift, Temperature, and Stress Effects 53
2.5 Takeaways 53
CHAPTER 3 Key Specifications..55
3.1 Why Specifications Matter.. 56
3.2 Core Measurement Specifications 59
3.2.1 Full-Scale Range (FSR) 59
3.2.2 Sensitivity... 61
3.2.3 Resolution 63
3.2.4 Noise Density 63
3.2.5 RMS Noise (Total Noise Over Bandwidth)...............65
3.2.6 Bandwidth (BW) and ODR 66
3.2.7 Difference between ODR and Bandwidth.................67
3.3 Accuracy and Error Sources 68
3.3.1 Axis Misalignment...69
3.3.2 Non-linearity...69

3.3.3 Zero-g Offset (Offset Error) 70
3.3.4 Offset Temperature Drift 70
3.3.5 Sensitivity Temperature Drift 71
3.3.6 Cross-Axis Sensitivity 73
3.3.7 Hysteresis 74
3.3.8 Bias Instability (Long-Term Stability) 75
3.3.9 Total Error Band 75
3.4 Environmental and Reliability Specifications 77
3.4.1 Operating Temperature Range 77
3.4.2 Shock Resistance 78
3.4.3 Vibration Resistance 78
3.4.4 Humidity and Environmental Sealing 79
3.4.5 Mechanical Stress Sensitivity 80
3.4.6 ESD Protection 80
3.5 Electrical and Operational Specifications 81
3.5.1 Current Consumption 81
3.5.2 Power Modes and Wake-on-Motion 82
3.5.3 Self-Test (Built-In Test Capability) 83
3.5.4 Self-Test Response Interpretation and Tolerance ... 84
3.5.5 Startup Time (Turn-On Time) 86
3.5.6 Supply Voltage and I/O Voltage 86
3.5.7 Output Format (Digital Resolution and Data Representation) 87
3.5.8 Filtering Options (Analog and Digital) 88
3.5.9 FIFO and Data Management 89
3.6 How to Evaluate an Accelerometer Using Its Specifications 89
3.6.1 Identify the Required Full-Scale Range (FSR) 90
3.6.2 Evaluate Noise and Resolution Together 90

3.6.3 Check Bandwidth and ODR 91
3.6.4 Match Bandwidth to the Application 91
3.6.5 Analyze Temperature Drift and Stability.................92
3.6.6 Consider Power Consumption and Power Modes....92
3.6.7 Check Non-Linearity and Cross-Axis Sensitivity 93
3.6.8 Verify Shock, Vibration, and Environmental Limits93
3.6.9 Look at System-Level Factors.................................94
3.6.10 Compare Sensors Using a Weighted Criteria Approach 94
3.6.11 Always Consider Calibration Needs 95
3.7 Takeaways 96
CHAPTER 4 How to Read a Datasheet 98
4.1 Understanding the Structure of a Datasheet 99
4.1.1 Device Overview and Key Features 99
4.1.2 Block Diagram: The Internal Architecture 101
4.1.3 Electrical Characteristics.......................................102
4.1.4 Register Map and Configuration Settings..............102
4.1.5 Mechanical and Package Information 104
4.1.6 Application Notes and Typical Performance Curves 105
4.1.7 Performance Characteristics (The Most Important Section) 106
4.2 Reading Performance Specifications in a Datasheet 106
4.2.1 Full-Scale Range (FSR) 107
4.2.2 Sensitivity..108
4.2.3 Resolution 109
4.2.4 Interpreting Noise and Bandwidth Specifications in Datasheets 110
4.3 Understanding Accuracy and Error Specifications in a Datasheet 111

4.3.1 Zero-g Offset (Offset Error) 111
4.3.2 Offset Temperature Drift.. 112
4.3.3 Sensitivity Tolerance and Sensitivity Drift 113
4.3.4 Non-linearity 114
4.3.5 Cross-Axis Sensitivity 115
4.3.6 Alignment Error (Axis Misalignment) 115
4.3.7 Noise Density and RMS Noise................................. 115
4.3.8 Hysteresis 116
4.3.9 Total Error Band (TEB) or Overall Accuracy 116
4.4 Interpreting Electrical and Operational Specifications .. 117
4.4.1 Supply Voltage (Vdd) and I/O Voltage ($VddIO$) 118
4.4.2 Current Consumption (Operating Modes) 119
4.4.3 Startup Time (Turn-On Time or Power-Up Time) 120
4.4.4 Communication Interface Specifications (I^2C / SPI) 121
4.4.5 Self-Test Behavior and Specifications 121
4.4.6 Interrupt Behavior and Threshold Registers 122
4.4.7 FIFO and Data Buffering Specifications 122
4.4.8 Filtering Options (Analog and Digital) 124
4.5 Understanding Environmental and Reliability Specifications 124
4.5.1 Operating Temperature Range 125
4.5.2 Shock Resistance 125
4.5.3 Vibration Resistance... 126
4.5.4 Humidity, Moisture, and Environmental Sealing ..127
4.5.5 Mechanical Stress Sensitivity 128
4.5.6 ESD Ratings (HBM, CDM).................................... 128
4.5.7 Reliability Validation and Endurance Tests 130
4.6 From Raw Output to Final Acceleration Estimate.......... 130

4.7 Takeaways 133
CHAPTER 5 Common Mistakes with Accelerometers.............136
5.1 Choosing the Wrong Full-Scale Range.........................137
5.2 Misunderstanding Noise 142
5.3 Power Mode Misuse 151
5.4 Orientation and Axis Alignment Errors 155
5.5 Inadequate Filtering Choices.......................................160
5.6 Failing to Prevent Aliasing...164
5.7 Sensor Placement and Mechanical Stress 175
5.8 Misinterpreting Motion Patterns..................................179
5.9 Misusing Thresholds and Interrupts 183
5.10 Mistaking Noise Spikes for Motion.............................187
5.11 Mistaking Offset Drift for Failure...............................190
5.12 Treating Accelerometers in Isolation 191
5.13 Startup and Initialization Errors.................................194
5.14 Misusing Self-Test 197
5.15 Overreliance on Raw Data 199
5.16 Timing and Synchronization Errors 201
5.17 Takeaways..204
CHAPTER 6 Accelerometer Applications 207
6.1 Consumer Devices 209
6.1.1 Smartphones and Tablets 209
6.1.2 Wearables and Fitness Trackers 211
6.1.3 Laptops and Portable Computers 212
6.1.4 Cameras and Handheld Video Devices..................213
6.1.5 Gaming, Virtual Reality (VR), and Augmented Reality (AR) 213

6.1.6 Smart Home Devices 214
6.1.7 Household Appliances 215
6.1.8 Tools and Power Equipment 216
6.1.9 Home Safety and Security Devices 217
6.1.10 Toys, Remote Controls, and Consumer Gadgets .. 218
6.2 Automotive Systems 218
6.2.1 Airbag and Crash Detection Systems 218
6.2.2 Electronic Stability Control and Rollover Detection 220
6.2.3 Inertial Navigation and Dead-Reckoning 220
6.2.4 Ride Comfort and Suspension Systems 221
6.2.5 Electric Vehicle Motor and Drivetrain Monitoring 221
6.2.6 Vehicle Security, Impact, and Tilt Detection 222
6.3 Medical and Healthcare Devices 223
6.3.1 Fall Detection and Emergency Alert Systems 225
6.3.2 Sleep Monitoring and Circadian Rhythm Analysis 225
6.3.3 Activity Monitoring and Gait Analysis 226
6.3.4 Respiratory and Cardiac Micro-Movement Sensing 227
6.3.5 Tremor Detection and Movement Disorders 227
6.3.6 Rehabilitation and Physical Therapy Devices........228
6.3.7 Medication Adherence and Smart Drug Delivery ..230
6.4 Industrial and Manufacturing Equipment 230
6.4.1 Vibration Monitoring and Predictive Maintenance 231
6.4.2 Motor Health and Machinery Diagnostics............232
6.4.3 Robotics and Industrial Automation 233
6.4.4 Heavy Machinery and Mobile Equipment 233
6.4.5 Structural Health Monitoring 234
6.4.6 Conveyors, Production Lines, and Assembly Systems 235

6.5 Aerospace, Drones, and Robotics 236
6.5.1 Drone Stabilization and Flight Control...................236
6.5.2 Robotics: Motion Control and Vibration Compensation 238
6.5.3 Aerospace and Aviation Systems 239
6.5.4 Autonomous Navigation and Inertial Sensing 240
6.5.5 Gimbals, Stabilized Cameras, and Precision Payloads 240
6.6 Takeaways 241
CHAPTER 7 Current and Emerging Trends........................... 243
7.1 Ultra-Low Power Motion Sensors 243
7.1.1 Smarter Power Modes... 244
7.1.2 Integrated Event Detection to Reduce System Power 245
7.1.3 Duty-Cycling and Context-Aware Sampling 247
7.1.4 Applications Driving Ultra-Low Power Development 247
7.2 AI-Enabled Motion Detection..248
7.2.1 AI for Motion Classification and Noise Robustness 249
7.2.2 On-Sensor Machine Learning.............................. 249
7.2.3 Energy-Efficient Models for Edge and On-Sensor AI 250
7.2.4 Context-Aware Motion Interpretation 251
7.3 Integrated Inertial Sensor Architectures 253
7.3.1 Embedded Sensor Fusion Engines.........................254
7.3.2 Integration of Multiple Sensors in Compact Packages 257
7.4 Miniaturization and MEMS Integration 260

7.5 Emerging Accelerometer Applications 263
7.5.1 Advanced Wearable Health Monitoring 263
7.5.2 Trend Toward Edge-Based Industrial Condition Monitoring 264
7.5.3 Robotics and Autonomous Systems 265
7.5.4 Smart Home and Appliance Intelligence 265
7.5.5 AR, VR, and Human–Machine Interfaces.............. 266
7.5.6 Environmental and Structural Monitoring 266
7.6 Takeaways 268
Quick Reference Summary... 270
Glossary 275
Suggested Reading & Resources 286
About the Author 289

Introduction

Accelerometers are small sensors that measure acceleration, vibration, and the pull of earth's gravity. These three physical quantities may sound simple, but they allow a device to understand how it is moving, how it is tilted, and what is happening around it. Every time your phone rotates its screen, your watch counts your steps, your car senses a collision, or your laptop protects itself during free fall, an accelerometer is typically working quietly in the background.

Over the past two decades, accelerometers have become one of the most widely used sensors in modern electronics. They have moved from specialized equipment into phones, wearables, appliances, toys, vehicles, industrial systems, and medical devices. Their rise is not accidental. Advances in micrometer-scale mechanical structures, improvements in manufacturing techniques, and better integration with digital electronics have all contributed to making accelerometers an important part of many devices and systems.

What Accelerometers Are

It is important not to confuse accelerometers with simple "motion sensors" used in home lighting or security systems. Those sensors merely detect that something moved in front of them. Accelerometers provide a much deeper understanding because they measure motion itself as a physical quantity, including how fast an object speeds up or slows down, how it vibrates, and how gravity acts on it. This makes accelerometers far more informative and suitable for precision tasks.

Why Accelerometers Became So Widely Used

Accelerometers have existed for many decades, but the versions we rely on today became possible only with the development of MEMS (Micro-Electro-Mechanical-Systems) technology. MEMS made it possible to shrink the mechanical parts of an accelerometer and place them directly on a silicon chip, allowing the sensor to fit into small consumer devices.

Why Accelerometers Matter

Accelerometers matter because they allow electronic devices to sense and respond to the physical world. Without them, many everyday features would not exist, from screen rotation and step counting to collision detection and flight stabilization. By converting physical forces into usable information, accelerometers make modern devices more responsive, intuitive, and capable.

Accelerometers make these experiences possible by sensing changes in motion and orientation and producing signals that electronic systems can interpret. By providing a continuous measure of how a device moves, tilts, or vibrates, accelerometers enable software to make informed decisions about behavior and response.

How This Book Is Structured

This book is designed for readers who want to understand accelerometers without diving into heavy mathematics or complicated engineering concepts. The chapters follow a natural progression, beginning with the basic ideas of what accelerometers are and how they work. From there, the book explores key specifications, common applications, real-world examples, and practical design considerations. Each chapter builds on the previous one, while individual sections provide enough context for readers to focus on specific topics without needing to read the book cover to cover.

You can read the book from start to finish or move directly to the chapters that interest you most. Many ideas become clearer when you see them visually, so take your time with the illustrations and examples. The glossary is always available if you come across new terminology, and the quick reference summary at the end provides a fast way to review important concepts.

Throughout this book, the terms software, firmware, code, algorithm, and application are used in a general sense to describe signal processing, calibration, compensation, and control logic applied to accelerometer data. Unless explicitly stated otherwise, these terms refer broadly to functionality that may be implemented in embedded firmware on a microcontroller, in device drivers, or in higher-level application software running on a host processor. The specific implementation location depends on system architecture and design choices and does not affect the underlying concepts discussed here.

About the Figures

Unless otherwise stated, figures in this book are conceptual illustrations intended to convey principles and design considerations. Actual sensor implementations, performance, and behavior depend on device design, configuration, and operating conditions; datasheets and related manufacturer publications should be consulted for definitive specifications.

CHAPTER 1
Accelerometer Fundamentals

Accelerometers may be small, but they give electronic devices the ability to sense something fundamental: how inertial effects act on a device as it moves, tilts, vibrates, or falls. Understanding what an accelerometer actually measures is the first step toward understanding how these sensors work and why they have become essential in so many products. Before exploring the technologies behind accelerometers or the details of how they are designed, it is helpful to build a clear picture of the physical quantities these devices detect and why those quantities are so informative.

1.1 What Does It Mean to Measure Acceleration

When we say a device "measures acceleration," we are really talking about measuring how the motion of an object changes over time. Acceleration is a measure of how quickly the speed of an object changes. If an object speeds up, slows down, changes direction, falls, vibrates, or tilts under the influence of gravity, it is experiencing acceleration. An accelerometer senses these changes and turns them into electrical signals that electronics can process.

Acceleration appears throughout daily life whenever motion changes. You feel it each time a car starts moving from a stoplight or when an elevator begins to rise. You feel it in turbulence on an airplane or when you toss your phone onto the couch. Any time an object changes its motion in any way, acceleration is involved. This makes acceleration one of the most fundamental physical quantities we can measure.

To make this idea clearer, imagine standing still. Even when you are not moving, you are experiencing the acceleration of Earth's gravity. Gravity constantly pulls downward with a force that creates a measurable acceleration of roughly $9.8\ m/s^2$. This means accelerometers can sense gravity even when nothing else is happening. If you tilt an accelerometer slowly, you do not create motion in the usual sense of moving forward or backward, but you change how gravity pulls on the accelerometer's internal structure, and the accelerometer senses this change.

This is why accelerometers can do more than detect motion. They can also be used to infer tilt and orientation, such as whether a device is upright, sideways, or upside down, based on how gravity is sensed along each axis. Many everyday features rely on this simple ability. When your phone rotates its screen as you turn it, the accelerometer does not detect motion through space, it measures how gravity acts on the sensor as the phone changes its angle.

Acceleration also shows up in more dynamic ways. When an object vibrates, the back-and-forth motion produces continuously changing acceleration. The stronger the vibration, the larger the change in acceleration. This is why accelerometers are widely used in tools, machinery, and industrial equipment to monitor mechanical health. A machine with a worn-out bearing, for example, produces a different vibration pattern than a healthy one. By sensing these patterns, accelerometers help detect early signs of malfunction before more serious damage occurs.

Another way to understand acceleration is through a simple example, dropping an object. If you drop your phone by accident, it begins to fall under the influence of gravity. During this brief period of free fall, the accelerometer senses little to no acceleration because the sensor and its internal mass are falling together. When the phone hits something, the sudden stop produces a sharp acceleration spike. Modern phones use this characteristic pattern, a moment of near zero acceleration followed by a strong impact, to recognize a fall. This information can trigger protective mechanisms to help safeguard sensitive internal components, such as camera modules, and can also be logged to assess whether a device has experienced impacts that may affect reliability or fall outside normal usage conditions. Another example comes from some laptops with mechanical hard drives, where accelerometer readings were used to quickly move the drive heads to a safe position during a fall.

Acceleration can also occur when nothing visibly moves. Think about a person sitting in a car that suddenly takes off. For a brief moment, the person feels pushed back into the seat. Even though the person is not moving relative to the car interior, their body experiences acceleration because the car is speeding up. An accelerometer in the car would sense this change even if it could not "see" anything happening.

Because acceleration appears in so many forms, including gravity, vibration, impacts, changes in speed, and changes in direction, we can use accelerometers for a wide range of tasks. They help determine how a device is positioned, whether it is moving smoothly or roughly, and how forces act on it in different situations.

Acceleration is such a rich physical concept that different technologies have been developed over the years to measure it. Although these technologies operate differently, they all rely on the same basic idea: acceleration produces a physical change, such as motion of a mass, deformation of a material, or a change in an electrical property, that can be measured and converted into an electrical signal.

To visualize the idea, imagine a simple drawing of a block attached to a spring, as shown in **Figure 1.1**. When the block moves right or left, the spring stretches or compresses. The spring's stretch represents the motion, but the rate at which that stretch changes over time represents the acceleration. This simple analogy captures the essence of what accelerometers are trying to detect: tiny changes in movement that happen continuously and often invisibly.

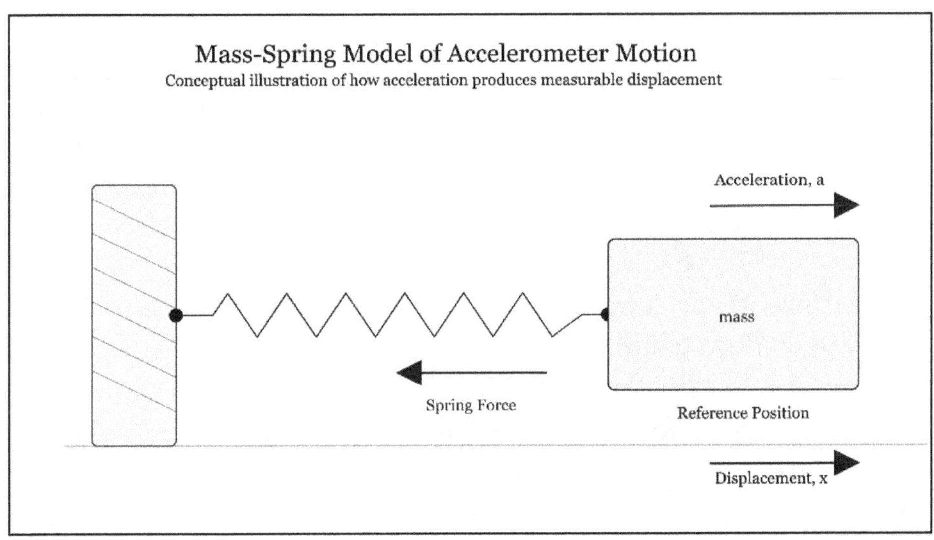

Figure 1.1 Simple mass spring model illustrating acceleration.

Acceleration is not a rare or exotic phenomenon. It is part of everyday life. That is what makes accelerometers so versatile and so valuable in the devices and machines we interact with every day.

Understanding acceleration and its many forms gives us a useful foundation for the chapters that follow. Different types of accelerometers rely on different technologies, and certain designs, especially capacitive MEMS accelerometers, have become especially common in modern products. But before diving into those details, it is helpful to look at the main types of accelerometers and the different physical principles they use to measure acceleration.

1.2 Types of Accelerometers

Accelerometers come in many forms, and each one measures acceleration in its own way. Although all accelerometers share the same goal, detecting how an object's motion changes over time, the physical principles behind them can be very different. Understanding the main types of accelerometers helps us appreciate the range of sensing technologies that exist today and shows why certain types are better suited for specific applications.

Different accelerometer technologies were developed over many decades. Some rely on mechanical motion, others on changes in electrical properties, and some depend on thermal effects or feedback forces. Before focusing on capacitive MEMS accelerometers, which dominate modern products, it is helpful to explore this broader landscape of how acceleration can be measured.

Below is an overview of the most common types of accelerometers, along with simple explanations and examples that make their operating principles easier to visualize.

1.2.1 Mechanical (Mass-Spring) Accelerometers

Mechanical accelerometers are the earliest form of acceleration sensors. They consist of a small mass attached to a spring inside a housing. When the accelerometer moves, the mass lags slightly behind because of inertia, causing the spring to stretch or compress. The amount of stretch provides a measurable indication from which the acceleration acting on the device can be determined.

These devices are intuitive to understand but tend to be larger and slower than modern sensors. Their simplicity made them useful for early automotive and industrial systems, but they are rarely used in compact electronic products today.

As illustrated earlier using the mass-spring analogy in **Figure 1.1**, mechanical accelerometers rely on the inertia of a small mass attached to a spring to sense acceleration.

The majority of accelerometer technologies are rooted in the mass-spring principle, where acceleration causes a mass to move relative to its housing. Even in modern sensors, this basic mechanical response is often the starting point, with the resulting motion later converted into an electrical signal using different sensing mechanisms.

1.2.2 Capacitive Accelerometers (Most Common in MEMS)

In capacitive accelerometers, acceleration causes a small internal mass to move slightly, changing the overlap or spacing between tiny capacitor plates. Because capacitance depends on both plate area and separation, even extremely small movements, sometimes only a few nanometers, produce measurable electrical changes.

Capacitive accelerometers offer excellent low frequency and static performance, low noise, low power consumption, suitability for high volume manufacturing, and good linearity and stability.

As a result, capacitive sensing has become the most widely used principle in modern MEMS accelerometers. By converting minute mechanical displacements into changes in capacitance, this approach enables accurate measurement of both dynamic and static acceleration, including gravity. We will explore capacitive MEMS accelerometers in much more detail in later sections and chapters.

1.2.3 Piezoelectric Accelerometers

Piezoelectric accelerometers use special materials, such as quartz or certain ceramics, that generate electric charge when compressed or stretched. In a piezoelectric accelerometer, a small mass presses on a piezoelectric crystal when the sensor experiences acceleration, creating a voltage proportional to the applied acceleration.

These accelerometers excel at measuring vibration and high frequency motion, which is why they are widely used in industrial machinery, aerospace testing, and structural monitoring. However, they are not ideal for measuring slow movements or static acceleration such as gravity, which limits their use in consumer products. Many piezoelectric accelerometers are MEMS based, using thin film piezoelectric layers to sense strain.

1.2.4 Piezoresistive Accelerometers

Piezoresistive materials change their electrical resistance when they are stretched or compressed. In a piezoresistive accelerometer, the proof mass (the small internal sensing mass) moves under acceleration and causes tiny deformations in beams or resistive elements. This deformation changes the electrical resistance, and that change is converted into a measurable electrical signal.

These devices are strong candidates for high-g shock measurements, which makes them useful in crash testing, automotive safety systems, and industrial environments. Like piezoelectric devices, piezoresistive accelerometers can also be manufactured using MEMS techniques.

1.2.5 Thermal (Convection) Accelerometers

Thermal accelerometers contain a tiny heater that warms a pocket of gas inside the sensor. Temperature sensors around the heater detect how the heat distribution changes when the device experiences acceleration. The shifting heat pattern reveals the direction and magnitude of acceleration.

These accelerometers have no mechanical moving parts, which makes them robust and durable. However, they generally have lower bandwidth and precision than capacitive devices, so they are used mainly in specialized applications where mechanical wear is a concern, such as long-life industrial monitoring or harsh-environment sensing.

1.2.6 Servo (Force-Balance) Accelerometers

Servo accelerometers, also known as force-balance accelerometers, operate using a closed loop feedback system. When the proof mass begins to move under acceleration, a controlled electrostatic or electromagnetic force is applied to push it back toward its neutral position. The magnitude of the feedback force required to keep the mass centered is proportional to the acceleration acting on the device.

Because the proof mass is held nearly stationary, these accelerometers offer extremely high accuracy and long-term stability. They are widely used in navigation systems, seismic measurement, and precision instrumentation. However, their increased circuit complexity, larger physical size, and higher cost make them less common in consumer electronics.

1.2.7 Resonant Accelerometers

Some accelerometers use vibrating structures, similar to tiny tuning forks. When acceleration is applied, the tension in these structures changes, altering their resonant frequency. By measuring this frequency shift, the acceleration acting on the device can be determined.

Resonant accelerometers can achieve excellent stability and very low noise, and some versions are manufactured using MEMS technology. They are used in precision applications such as inertial navigation and scientific instruments.

1.2.8 Optical Accelerometers

These accelerometers are highly sensitive and immune to electromagnetic interference, making them useful in scientific research and specialized environments. However, their complexity, size, and cost generally limit their use to niche and high-performance applications rather than everyday consumer products.

1.2.9 Emerging Accelerometer Technologies

Researchers continue to explore new approaches to acceleration sensing, including graphene-based devices, nanoscale resonators, quantum accelerometers, and advanced thermal and optical designs. These technologies show potential for future navigation, robotics, and high precision measurement.

At present, most of these approaches remain confined to research laboratories and specialized applications. Although they demonstrate impressive performance in controlled settings, they are not yet suitable for widespread, high-volume production or everyday consumer products.

1.3 MEMS Accelerometers

When people talk about "modern accelerometers," they usually mean MEMS accelerometers, but MEMS is not a sensing principle by itself. MEMS, short for Micro Electro Mechanical Systems, is a method of building tiny mechanical structures on or within a substrate, most commonly silicon. The key idea is that both mechanical elements, such as beams, springs, and proof masses, and electronic components can be created using techniques similar to semiconductor fabrication, with additional steps that shape and release mechanical parts.

MEMS is a manufacturing platform, not a measurement method. This distinction matters because multiple sensing technologies can be implemented using MEMS fabrication, including capacitive accelerometers, which are the most common today.

Even within the MEMS category, there is a wide variety of designs, geometries, and operating principles. What they all share is the ability to be made extremely small, lightweight, and highly repeatable at large volumes. This combination is what allowed accelerometers to move from specialized tools into mainstream consumer products.

Before the 1990s, accelerometers were typically bulky, expensive, and used mainly in aerospace, automotive crash testing, and industrial instrumentation. These devices worked well but were far too large and costly to fit inside a phone or wearable device.

MEMS technology changed that landscape by enabling accelerometers to be manufactured at the wafer level, dramatically reducing size, cost, and power consumption while improving consistency and scalability. This shift helped move accelerometers from specialized instruments into mainstream products.

1.3.1 Miniaturization

MEMS fabrication allows the core sensing structure of an accelerometer, including springs, beams, masses, and electrodes, to be formed on a silicon substrate with micrometer-level precision through a combination of material deposition, patterning, and selective etching. Structures that once required hand assembly under a microscope can now be created directly at the wafer level. This miniaturization is what enabled accelerometers to become small enough for handheld electronics and embedded systems.

At the structural level, a MEMS accelerometer is built around a few key mechanical elements that work together inside a very small footprint. The proof mass provides the moving element, the spring defines how it is suspended, the anchor fixes the structure to the substrate, and the sensing electrodes detect its motion.

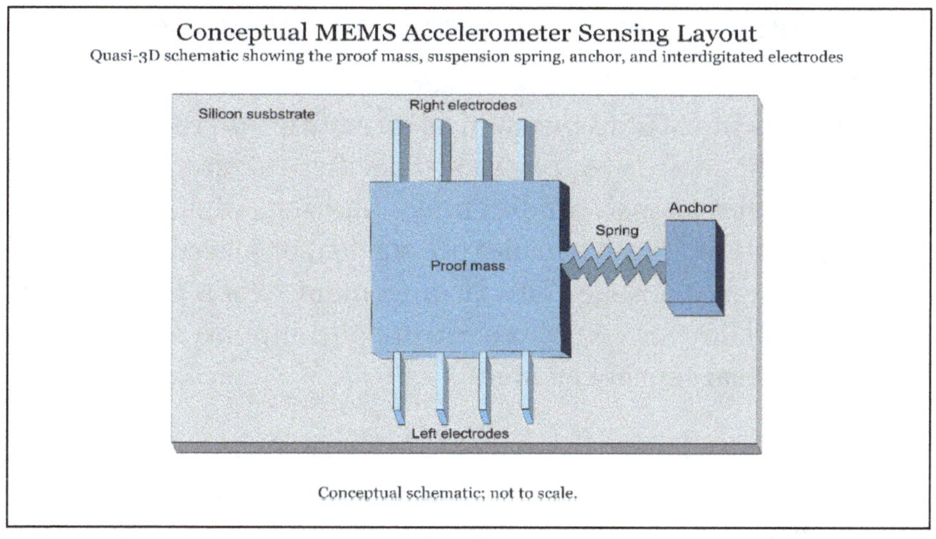

Figure 1.2 Simplified quasi-3D view of a MEMS accelerometer die highlighting the proof mass, suspension spring, fixed anchor, and sensing electrodes.

Figure 1.2 provides a simplified quasi-3D view of a MEMS accelerometer die, illustrating how the proof mass, suspension spring, anchor, and left and right sensing electrodes are formed on the silicon substrate.

1.3.2 High Volume Manufacturing

MEMS devices are produced using wafer-level processes, just like integrated circuits. Entire wafers containing hundreds or thousands of sensors can be processed simultaneously. After fabrication, individual dies are separated, bonded, and packaged. This high-volume workflow is what makes MEMS accelerometers affordable.

Because the mechanical structures are created lithographically rather than assembled manually, the resulting devices are highly consistent from unit to unit. This uniformity is essential for large-scale consumer and automotive markets.

1.3.3 Compatibility With Electronic Integration

Most MEMS accelerometers use a two-die architecture. One die contains the MEMS mechanical structure, and the second die contains the ASIC (Application-Specific Integrated Circuit), which handles signal conditioning, filtering, digitization, and communication. These two dies are stacked or placed side by side inside a single package. This arrangement keeps the footprint small while allowing complex electronics to support and interpret the mechanical sensing element.

The tight coupling between MEMS structures and ASICs is a key part of what makes modern accelerometers so powerful. The ASIC can amplify extremely small signals, reduce noise, apply filters, manage calibration, track temperature effects, and even implement built-in features such as tap detection or step counting.

1.3.4 Low Power Consumption

MEMS accelerometers are engineered for extremely low power operation, with some devices consuming only a few μA (microamperes) in low power modes, making them well suited for smartphones, wearables, medical patches, wireless IoT sensors, and other battery-operated equipment. Some accelerometers also include special low power wakeup engines that run continuously at microampere current levels and alert the main processor only when needed.

1.3.5 High Reliability and Robustness

Despite their tiny size, MEMS accelerometers are highly durable. Their mechanical structures are firmly anchored into the silicon substrate and are designed to withstand shock, vibration, repeated impacts, and temperature changes. This robustness is why MEMS accelerometers operate reliably in environments ranging from smartphones to automotive systems and industrial machinery.

1.3.6 Why the Industry Chose Capacitive MEMS

Although MEMS supports multiple sensing principles, capacitive MEMS accelerometers became the dominant choice for most consumer and general-purpose applications. This was driven by the advantages discussed above, along with good stability over temperature and excellent manufacturability at high volume and low cost. The next section looks more closely at capacitive MEMS accelerometers and the reasons they became so widely used.

1.4 Capacitive MEMS Accelerometers

Capacitive MEMS accelerometers are the most widely used accelerometers in the world today. They are found in smartphones, wearables, drones, cars, medical devices, gaming controllers, industrial machinery, and many battery-powered devices that need to sense motion or orientation. Their widespread adoption is not accidental. They offer a balanced combination of small size, high sensitivity, stability, low noise, low power consumption, reasonable cost, and efficient manufacturability.

To understand why this technology became the dominant solution, it is helpful to examine two aspects. The first is how capacitive sensing works inside a MEMS device. The second is which characteristics make this approach especially well-suited for high volume applications.

1.4.1 The Principle of Variable Capacitance

Capacitive MEMS accelerometers measure motion by detecting small changes in electrical capacitance between conductive elements. When the proof mass moves under acceleration, the spacing between these elements changes, modifying the electric field and therefore the capacitance. Although these changes are extremely small, they can be measured with high precision and form the basis of many modern accelerometer designs.

When acceleration causes the proof mass to move, the relative spacing between the movable and fixed plates changes. This produces a small change in capacitance, which is detected by the sensor electronics and used to determine the applied acceleration, as illustrated in **Figure 1.3**.

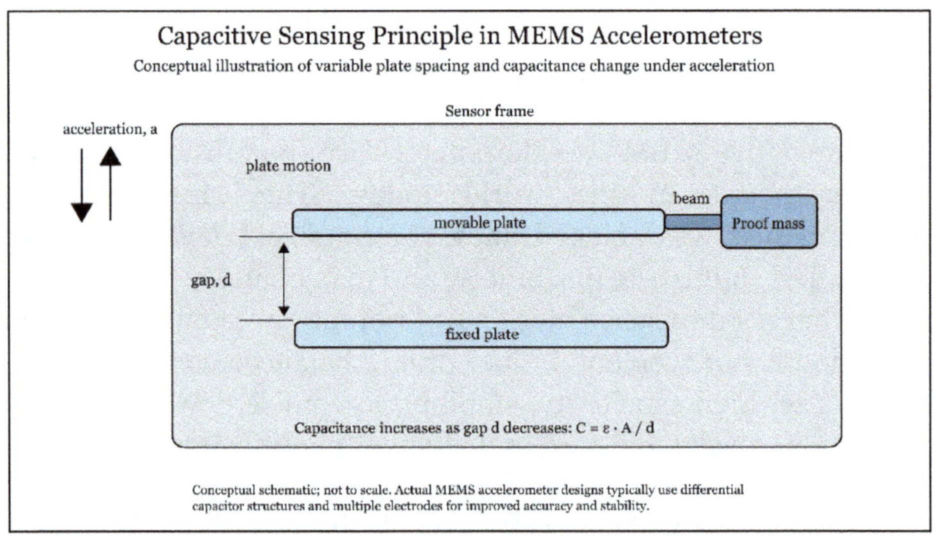

Figure 1.3 Capacitive sensing principle showing two parallel plates with changing spacing under acceleration.

Capacitive accelerometers rely on a simple but powerful concept. When two conductive plates are positioned close to each other, the electrical capacitance between them changes if the distance between the plates or the amount of overlap changes.

Mathematically, capacitance is given by:

$$C = \epsilon \cdot \frac{A}{d}$$ ***Equation 1.1***

where C is the capacitance, ε is the permittivity of the material between the plates, A is the overlapping area of the plates, and d is the distance between them.

In the context of MEMS accelerometers, both A and d are extremely small. Changes of only a few nanometers (nm) in the plate spacing can produce measurable changes in capacitance. MEMS fabrication techniques allow these structures to be manufactured with micrometer (μm)-level precision and with very small initial gaps.

1.4.2 Differential Capacitive Sensing: The Key to Accuracy

Most capacitive MEMS accelerometers use differential capacitance rather than a single capacitor. This approach significantly improves accuracy, noise immunity, and long-term stability.

Using multiple sensing fingers in parallel increases the overall capacitance change and makes small proof-mass motion easier to detect. This helps improve sensitivity and supports more stable measurement of very small displacements.

A dedicated ASIC measures the difference between these two capacitances, which can be expressed as:

$$\Delta C = C_1 - C_2$$ ***Equation 1.2***

This differential sensing strategy is a key reason why capacitive MEMS accelerometers achieve high stability, excellent noise performance, and reliable operation across a wide range of conditions.

Figure 1.4 illustrates differential capacitive sensing using interdigitated comb fingers, where $C1$ and $C2$ represent the total capacitance of parallel finger arrays on opposite sides of the proof mass. When acceleration causes the proof mass to move vertically, the gap between one set of movable and fixed fingers decreases while the gap on the opposite side increases. This opposing change in plate spacing produces a differential capacitance output, as expressed in $\boldsymbol{\Delta C} = C_1 - C_2$.

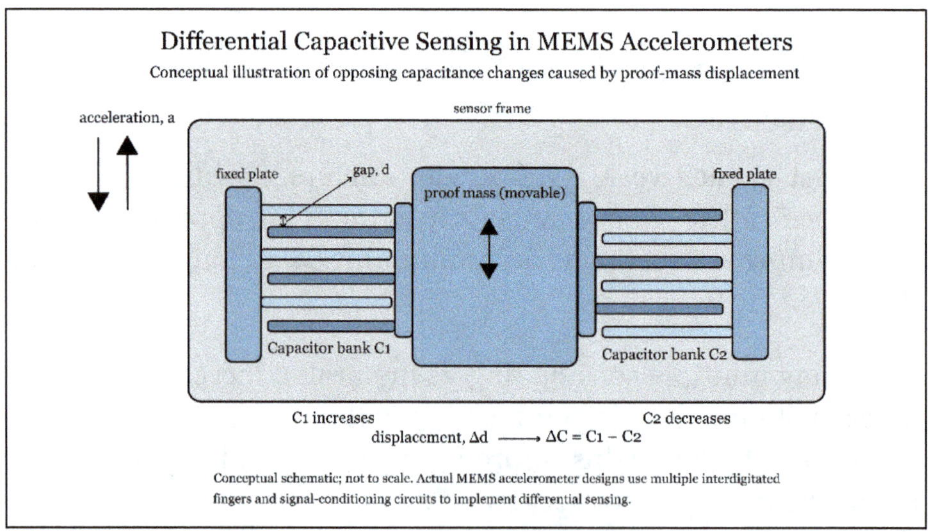

Figure 1.4 Differential comb-finger capacitive sensing with opposing capacitance changes (C1 - C2) caused by proof-mass displacement.

1.4.3 Why MEMS Makes Capacitive Sensing Possible

Capacitive sensing works best when the distance between capacitor plates is extremely small. MEMS fabrication makes this possible by enabling gaps of only a few micrometers (μm), beams and springs with highly consistent dimensions, and precise, repeatable capacitor geometries. It also allows large numbers of capacitors to be integrated into a compact area. The ability to mass produce these structures on silicon wafers is what makes capacitive MEMS accelerometers affordable, uniform, and scalable across millions of units.

1.5 A Simple Mathematical Intuition Behind Acceleration

Up to this point, we have looked at acceleration in a largely intuitive way. We described how objects change their motion, how gravity acts as a constant acceleration, and how a MEMS accelerometer senses tiny mechanical shifts. In this section, we take a closer look at the basic mathematics behind acceleration. The goal is not to turn this book into a physics course, but to give you enough mathematical intuition to understand how accelerometers interpret the physical world.

Acceleration is one of the simplest physical concepts, yet also one of the most powerful. It tells us how quickly motion changes. Unlike velocity, which describes how fast something moves, acceleration describes how fast the speed changes. Even if the mathematics is minimal, having a clear understanding of this relationship helps explain everything that happens inside an accelerometer.

1.5.1 Acceleration as the Change of Velocity Over Time

Acceleration is defined mathematically as:

$$a = \frac{\Delta v}{\Delta t} \qquad \textbf{\textit{Equation 1.3}}$$

where a is acceleration, Δv is the change in velocity, and Δt is the time interval over which the change occurs.

If a car speeds up from 0 to 10 meters per second in 2 seconds, the acceleration is:

$a = (10 - 0) / 2 = 5\, m/s^2$ ***Equation 1.4***

This means the car's velocity increases by 5 meters per second every second.

Accelerometers detect this change directly. They do not need to know the object's speed or direction, only how quickly the motion changes over time.

1.5.2 Gravity as a Constant Acceleration

One of the most important ideas for understanding accelerometers is that gravity is an acceleration, not a force unique to Earth. Gravity accelerates all objects downward at approximately:

$g = 9.81\, m/s^2$ ***Equation 1.5***

This means that even when you are standing still, an accelerometer senses gravity as a constant acceleration. This is why accelerometers can detect tilt. When the sensor is rotated slowly, the direction of gravity relative to the sensor changes, and the accelerometer measures this change even though the device is not moving through space.

This also explains why accelerometers output *zero g* during free fall. When an object is falling, it is no longer resisting gravity. The internal proof mass and the sensor housing accelerate together, so the accelerometer momentarily senses no acceleration at all.

1.5.3 Relating Acceleration to Proof Mass Displacement

For a MEMS accelerometer, acceleration causes a change in the position of the proof mass. A simplified mass-spring system follows the relation:

$$F = m\,a = k\,x$$ ***Equation 1.6***

where m is the mass, a is acceleration, k is the spring constant, and x is the displacement of the mass.

Rearranging this expression gives:

$$x = (m\,/\,k)\,a$$ ***Equation 1.7***

This simple relationship explains how inertia-based sensors work. When acceleration occurs, the mass moves, and the amount of displacement depends on the proof mass, the stiffness of the springs, and the magnitude of the applied acceleration. A larger mass or softer springs increase sensitivity, while a lighter mass or stiffer springs reduce sensitivity. This behavior is illustrated conceptually in **Figure 1.1**

As a simple example, consider a proof mass of 2 micrograms, suspended by springs with an effective stiffness of 0.2 N/m. Under an acceleration of 1 g, or 9.81 m/s^2, $\boldsymbol{x}$ = $(m\,/\,k)\,a$ **Equation 1.7** gives:

$$x = (m\,a)\,/\,k$$

$$x = \frac{(2\times10^{-9})(9.81)}{0.2}$$ ***Equation 1.8***

$$x \approx 9.8 \times 10^{-8}\,m$$ ***Equation 1.9***

This displacement is on the order of 100 nm, far smaller than the width of a human hair, yet well within the range that MEMS capacitive structures can detect. This is one of the reasons MEMS technology is so effective: extremely small physical movements can be converted into meaningful electrical signals.

1.5.4 Relating Tilt to Acceleration

Because gravity is a constant acceleration vector, an accelerometer measures different values depending on its orientation relative to gravity. When a sensor is tilted by an angle θ, the component of gravitational acceleration measured along the horizontal axis is $g\ cos(\theta)$, while the component measured along the perpendicular axis is $g\ sin(\theta)$. This is why tilt can be measured even when the device is not moving. **Figure 1.5** illustrates this projection of the gravity vector onto the sensor axes as the device is tilted.

A simple tilt example helps illustrate this behavior. If a sensor is rotated by 30°, the acceleration measured along the horizontal axis is:

$$a_{\mathrm{X}} = g\ cos(30°) \approx 0.87\ g \qquad \textit{Equation 1.10}$$

and along the perpendicular axis:

$$a_{\mathrm{Y}} = g\ sin(30°) = 0.5\ g \qquad \textit{Equation 1.11}$$

The equations above describe the magnitudes of the projected gravity components; the sign of each measured value depends on the orientation and polarity of the sensor axes relative to gravity, as shown in **Figure 1.5**.

These predictable relationships allow devices to determine whether they are upright, sideways, upside down, or tilted by a known angle. A smartphone uses this information to rotate the screen, a drone uses it to maintain balance, and a medical device can use it to estimate patient posture.

Figure 1.5 illustrates how the signs of the measured acceleration components depend on the orientation of the sensor axes. In this convention, gravity acts downward in the world frame, and the accelerometer reports the signed projections of the gravity vector onto its rotated sensor axes. The angle θ represents a counterclockwise rotation of the sensor axes relative to the world frame in the x-y plane.

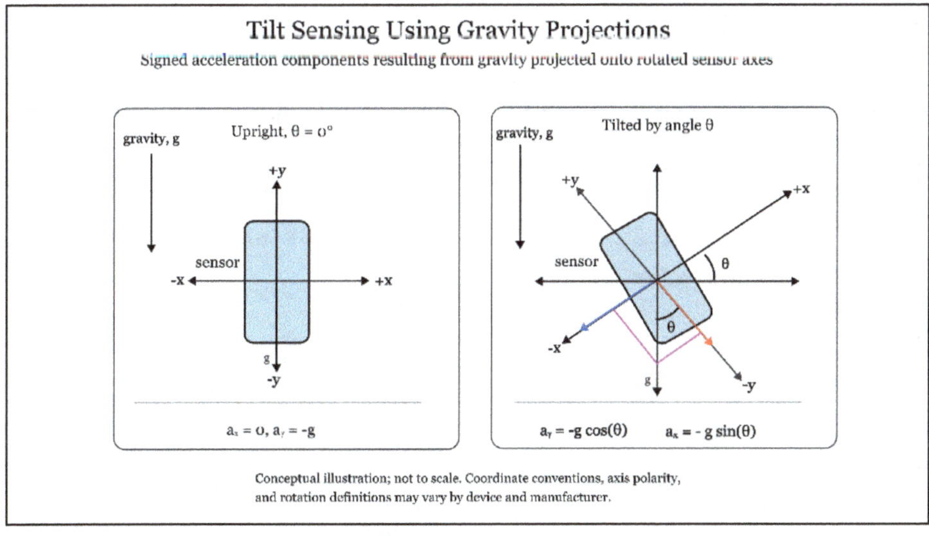

Figure 1.5 Tilt sensing using gravity projections onto rotated sensor axes, showing the signed acceleration components measured along the x and y directions.

1.5.5 Vibrations and Frequency Response

In vibrating systems, acceleration varies rapidly with time. For sinusoidal motion (for example, vibration at a single frequency), displacement ($x(t)$), velocity ($v(t)$), and acceleration($a(t)$) are related as follows:

$$x(t) = A sin(2\pi f t)$$

$$v(t) = A(2\pi f) cos(2\pi f t)$$

$$a(t) = -A(2\pi f)^2 sin(2\pi f t)$$ ***Equation 1.12***

where f is the frequency and A the amplitude.

This relationship highlights an important insight. High frequency vibrations produce large acceleration values even when the motion amplitude is small, while low frequency vibrations produce much smaller accelerations for the same amplitude, as illustrated in **Figure 1.6.**

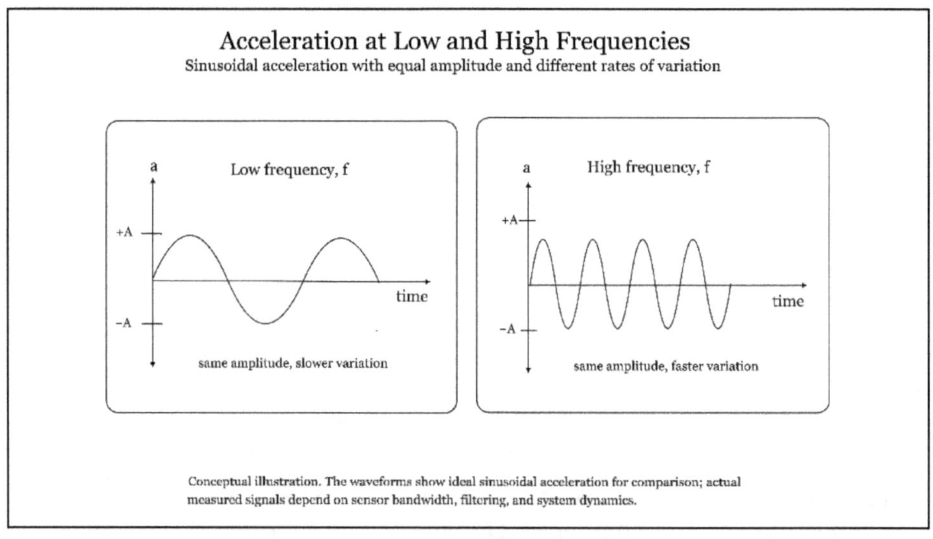

Figure 1.6 Sinusoidal acceleration at low and high frequencies with the same amplitude (±A).

1.6 Real-World Applications of Accelerometers

Accelerometers play a role in far more products than most people realize. Because they can sense motion, vibration, orientation, gravity, and sudden impacts, they act as a form of motion awareness within many modern systems. While the internal structures of accelerometers operate at microscopic scales, the functions they enable are familiar, practical, and often essential to product performance, safety, and user experience.

Across industries, accelerometers are rarely used in isolation. Instead, they serve as inputs to larger systems that interpret motion patterns, detect events, and make decisions based on acceleration data. The same basic sensing principles discussed earlier in this chapter appear repeatedly in very different contexts, with system requirements determining how measurements are filtered, interpreted, and acted upon.

In consumer electronics, accelerometers support intuitive interaction by enabling screen rotation, gesture recognition, activity tracking, and motion-based user interfaces. In vehicles, they contribute to safety, comfort, and diagnostics by detecting impacts, monitoring vehicle dynamics, and sensing vibration or tilt. Medical and healthcare devices rely on accelerometers to monitor movement, posture, and subtle motion over extended periods, enabling applications such as fall detection, rehabilitation tracking, and physiological monitoring.

Industrial and manufacturing environments use accelerometers primarily as diagnostic tools. By measuring vibration and motion patterns, these systems detect wear, imbalance, misalignment, and early signs of mechanical failure. In aerospace, drones, and robotic platforms, accelerometers form part of inertial sensing systems that support stabilization, navigation, and real-time control under dynamic conditions. Even in everyday household and embedded products, accelerometers quietly enhance usability, safety, and automation through motion-aware behavior.

These examples illustrate an important theme: accelerometers are valuable not simply because they measure acceleration, but because their outputs can be transformed into meaningful system behavior. The same sensor may serve very different roles depending on its operating environment, required accuracy, power constraints, and response time.

1.7 Takeaways

This chapter introduced the core ideas behind acceleration, outlined the main technologies used to measure it, and showed why capacitive MEMS accelerometers dominate modern products.

We began by exploring the basic meaning of acceleration and how accelerometers translate changes in motion into measurable signals, converting physical effects into information that electronics can interpret.

We then introduced the major sensing principles used in accelerometers. Mechanical mass-spring systems, piezoelectric and piezoresistive materials, capacitive structures, thermal sensors, force-balance mechanisms, and resonant designs represent different ways engineers have approached acceleration measurement. Together, they illustrate the range of solutions that have been developed to sense the same physical quantity, each with strengths and limitations depending on the application.

We then introduced MEMS accelerometers and explained why they made modern accelerometers small, reliable, and affordable. By allowing miniature sensor structures to be formed directly on silicon with microscopic precision, MEMS technology made it possible to mass produce consistent devices at low cost.

This helped accelerate the widespread use of accelerometers in consumer, automotive, medical, and industrial products.

We also looked inside a typical capacitive MEMS accelerometer to understand how its key components work together. The proof mass, suspension springs, comb fingers, and sensing electrodes collectively convert tiny physical motions into accurate digital signals. Even displacements on the scale of nanometers can carry meaningful information about acceleration.

Capacitive MEMS accelerometers emerged as the most widely used type because of their stability, low noise, low power consumption, and compatibility with high-volume manufacturing. Their strong performance in static and low frequency measurements makes them well suited for applications ranging from smartphones and medical devices to industrial machinery and automotive systems.

Finally, we introduced simple mathematical relationships that explain how accelerometers interpret motion, tilt, and vibration, and then connected those principles to real-world applications. Together, these ideas provide a solid foundation for understanding both the underlying behavior of accelerometers and their practical value in modern systems.

With this chapter, you now have a broad understanding of what accelerometers measure, the technologies that make measurement possible, and why these sensors are so widely used. In the next chapter, we look more closely inside MEMS accelerometers to see how their mechanical structures and electrical principles work together.

CHAPTER 2
Inside a MEMS Accelerometer

Accelerometers are often described in simple terms: they measure acceleration, vibration, and the pull of gravity. But how does a device small enough to fit inside a smartwatch or smartphone detect such subtle physical changes? What happens inside the tiny silicon structure when the device tilts, moves, or experiences a sudden jolt? And how is that motion converted into a digital number that software can interpret?

This chapter takes us inside the accelerometer itself. Building on the fundamental concepts introduced in Chapter 1, we now examine the internal mechanisms that allow a capacitive MEMS accelerometer to function. Although these sensors are extremely small, sometimes only a few millimeters across, they contain carefully engineered mechanical structures, electrodes, and signal processing circuits that work together with remarkable precision. Understanding how these components interact helps explain why accelerometers behave the way they do, why certain design choices matter, and how engineers optimize these devices for performance, reliability, and low power consumption.

The focus of this chapter is on capacitive MEMS accelerometers, the type most widely used in consumer, automotive, medical, and industrial applications. Rather than surveying every sensing technology, we concentrate on how this architecture works, from the motion of the internal proof mass, to the resulting change in capacitance, and finally to the analog and digital signal processing performed by the ASIC. Along the way, we introduce the key physical principles, simple mathematical relationships, and practical examples that make the concepts intuitive.

2.1 Core Principle: From Mechanical Motion to Electrical Signal

Accelerometers measure acceleration by converting mechanical motion into an electrical signal. When acceleration acts on the device, an internal proof mass moves slightly relative to the surrounding structure. This motion alters an electrical property inside the sensor, which is then processed and converted into a digital value that software can interpret.

2.1.1 From Mechanical Motion to Electrical Signals and Digital Data

A modern capacitive MEMS accelerometer operates across two interconnected domains: a mechanical domain and an electrical domain that includes both analog and digital processing.

In the mechanical domain, acceleration causes the proof mass to shift relative to the sensor frame. As shown in $\boldsymbol{x = (m / k)\, a}$, the magnitude of this displacement depends on the proof mass and the stiffness of the suspension springs. Even though the motion is extremely small, it is well controlled and repeatable.

In the electrical domain, this mechanical motion changes the spacing or overlap of internal capacitor electrodes. The resulting capacitance change is converted into a voltage by the ASIC, filtered, digitized by an analog-to-digital converter (ADC), and formatted into a digital output. This chain of transformations enables the sensor to convert physical motion into usable numerical data.

2.1.2 Why Mechanical Motion Is So Small, Yet So Useful

At first glance, it may seem surprising that accelerometers rely on physical motion at all, given how small MEMS structures are. However, MEMS fabrication allows this motion to be precisely controlled and highly repeatable across millions of devices.

Because the internal capacitors are also extremely small and operate in large parallel arrays, even nanometer-scale displacements produce measurable electrical changes. This combination of controlled motion and sensitive electrical detection is what makes capacitive MEMS accelerometers both practical and reliable.

2.1.3 Why Orientation Is Measurable Even Without Motion

A key characteristic of accelerometers is their ability to measure gravity as a constant acceleration. When the device is tilted, the direction of gravity relative to the sensor axes changes, producing a steady shift in the proof mass. This allows accelerometers to determine orientation even when the device is not moving.

2.2 Inside a Capacitive MEMS Accelerometer

To understand how a capacitive MEMS accelerometer works, it helps to picture the device not as a simple electronic chip, but as a miniature mechanical instrument carved into silicon. Although the sensor package may be only a few millimeters across, the internal components are arranged with extraordinary precision.

At the center of the device is the proof mass, a suspended silicon structure that moves slightly when acceleration is applied.

This motion alters the spacing or overlap of tiny capacitor plates, producing a measurable change in capacitance. This is the fundamental sensing mechanism of a capacitive MEMS accelerometer.

2.2.1 The Proof Mass: A Tiny Inertial Element

The proof mass is a small but critical part of the accelerometer. During acceleration, the mass briefly resists motion due to inertia, creating a relative displacement between the mass and the surrounding frame. Although this displacement is extremely small, often in the order of tens to hundreds of nanometers, it carries meaningful information about the applied acceleration. The location and role of the proof mass within a typical three-axis MEMS structure are illustrated in **Figure 2.1.**

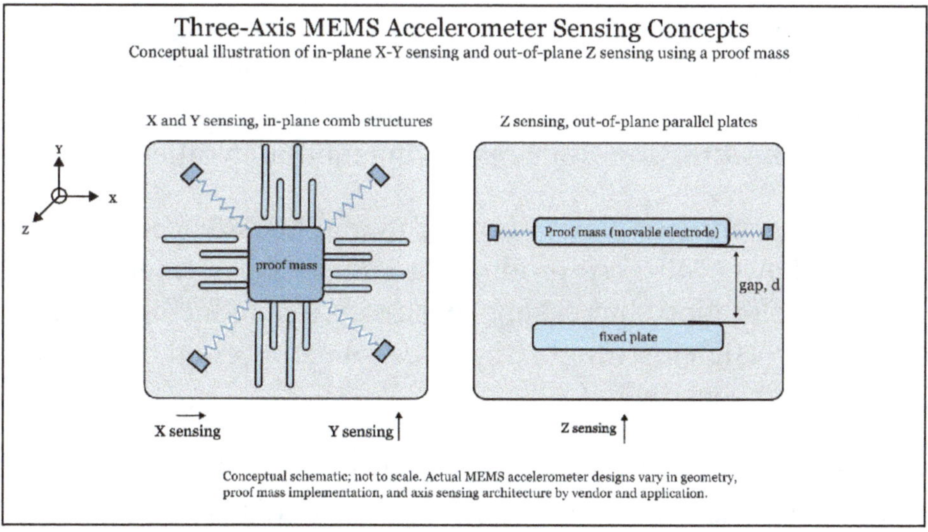

Figure 2.1 Conceptual illustration of three-axis accelerometer sensing, showing in-plane X–Y sensing structures and out-of-plane Z sensing using a proof mass.

Figure 2.1 illustrates two common sensing arrangements in a three-axis MEMS accelerometer. The *X* and *Y* axes are often measured using in-plane comb structures, while the *Z* axis is commonly measured using an out-of-plane parallel-plate structure. In both cases, the proof mass moves slightly under acceleration, and that motion is converted into a measurable change in capacitance.

The size and shape of the proof mass directly influence sensitivity, noise performance, resonant frequency, robustness against shock, and the direction or directions in which acceleration can be measured. In some designs, a single proof mass is used to sense multiple axes of acceleration, while in others, separate sensing structures are implemented for the X, Y, and Z axes.

2.2.2 Suspension Springs: Allowing Motion and Setting Sensitivity

The proof mass is connected to the surrounding frame by flexible suspension springs. These springs are thin silicon beams, typically a few micrometers wide, that allow controlled movement in one direction while restricting motion in others.

Spring stiffness plays a central role in MEMS accelerometer design. Lower stiffness produces a larger displacement for the same acceleration, increasing sensitivity but reducing resonant frequency. Higher stiffness improves robustness and bandwidth, but reduces sensitivity.

In practice, engineers must balance sensitivity against shock resistance, bandwidth against noise, and displacement range against linearity. Folded or serpentine spring geometries are commonly used because they provide compliant motion in the sensing direction while resisting movement in unwanted directions.

Figure 2.2 illustrates three common suspension spring layouts used in MEMS accelerometers for in-plane sensing. Folded-beam springs allow greater motion along the sensing direction while maintaining good symmetry and stress balance, which helps reduce cross-axis sensitivity. Serpentine springs achieve large motion within a compact footprint by increasing the effective spring length through multiple bends, making them well suited for designs that require low stiffness in limited area, such as compact consumer devices or space-constrained sensor layouts.

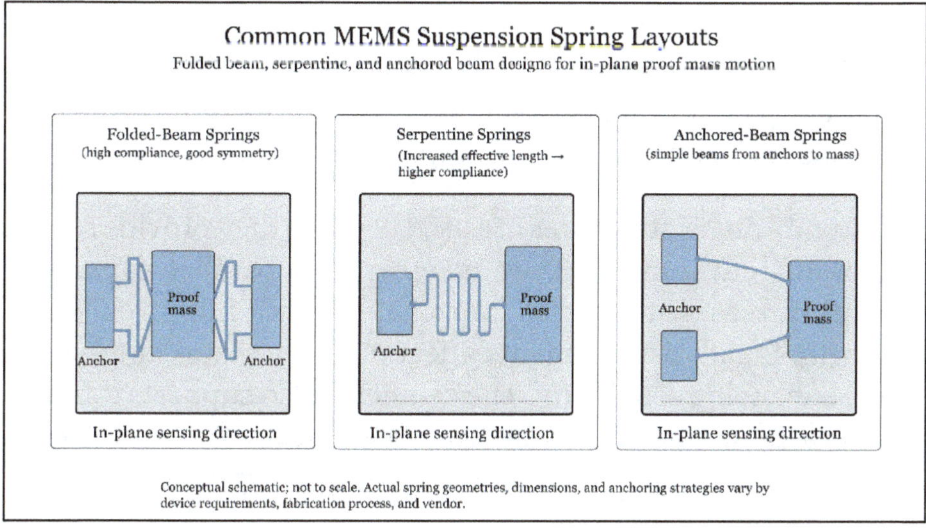

Figure 2.2 Examples of folded-beam, serpentine, and anchored-beam suspension springs used for in-plane proof-mass motion in MEMS accelerometers.

Anchored-beam springs use one or more bending beams connected between fixed anchors and the proof mass, allowing in-plane motion through elastic beam deflection with a relatively simple geometry. While the specific geometries vary by vendor and application, all three layouts serve the same fundamental role of suspending the proof mass and defining the mechanical sensitivity of the accelerometer along the chosen sensing axis.

2.2.3 Comb-Finger Structures: Capacitive Sensing

Capacitive MEMS accelerometers typically use interlocking comb fingers, where one set of fingers is attached to the proof mass and is movable, while the other set is anchored to the frame and remains fixed. As illustrated in **Figure 2.1** for the in-plane X and Y sensing axes, these comb structures are arranged so that relative motion occurs within the plane of the silicon die.

When the device accelerates, the proof mass shifts and the movable comb fingers slide relative to the fixed ones, changing the overlap or spacing between the capacitor plates. This motion creates a measurable change in capacitance.

Comb-finger structures are used because they provide several important advantages. They create many parallel capacitors, which increases the total capacitance and improves signal strength. Small displacements affect multiple plates at the same time, enhancing sensitivity. The geometry also supports differential measurement, improving accuracy, and helps reduce sensitivity to temperature drift and manufacturing variation.

The geometry of the comb fingers, including their length, width, spacing, and number, plays a major role in determining the overall performance of the accelerometer. For example, longer or more numerous comb fingers can increase capacitance and sensitivity, while spacing, width, and layout also affect mechanical behavior, parasitic capacitance, and manufacturability.

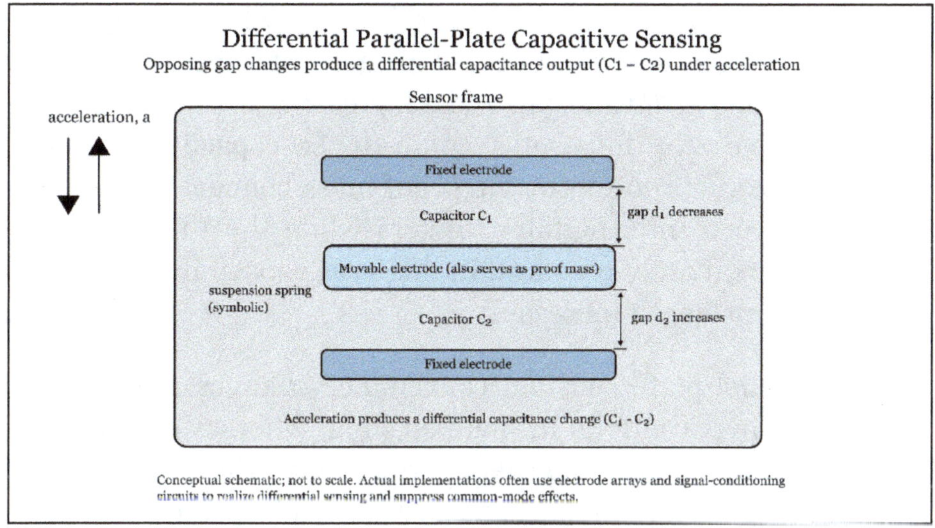

Figure 2.3 Differential parallel-plate capacitive sensing in a MEMS accelerometer. Applied acceleration causes the movable electrode to shift, decreasing gap d_1 and increasing gap d_2, resulting in a differential capacitance change (C_1 - C_2).

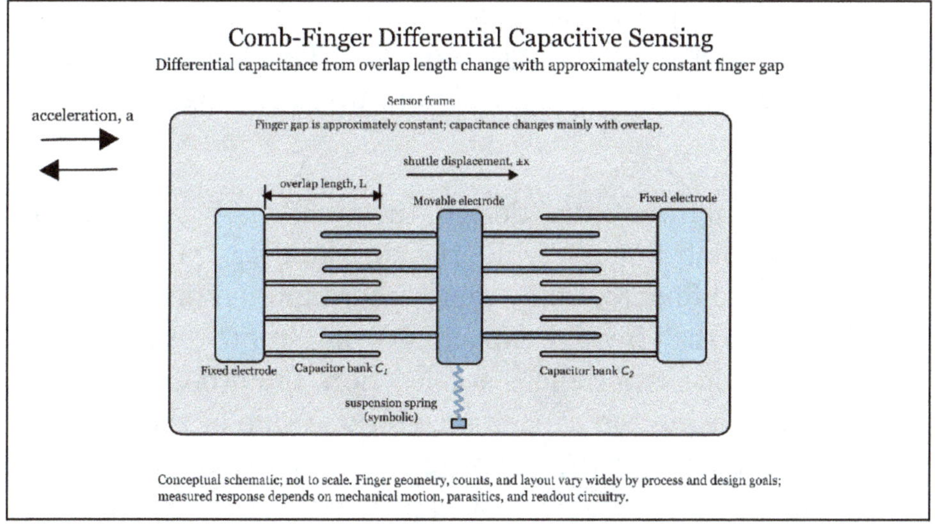

Figure 2.4 Comb-finger capacitive sensing in a MEMS accelerometer. Displacement of the proof mass changes the overlap length between interdigitated fingers while the finger gap remains approximately constant, producing a differential capacitance change.

Figure 2.3 illustrates differential parallel-plate capacitive sensing, where acceleration causes the movable electrode to shift, decreasing one electrode gap while increasing the other. In contrast, **Figure 2.4** illustrates comb-finger capacitive sensing, where horizontal motion of the proof mass changes the overlap length between interdigitated fingers while the finger gap remains approximately constant. In this case, capacitance increases as the overlap length increases.

In parallel-plate sensing, capacitance changes are typically dominated by variations in plate spacing, whereas comb-finger structures are designed so that changes in overlap length dominate the response, reducing sensitivity to gap variation.

2.2.4 Differential Capacitors: The Engine of High Precision

Most capacitive MEMS accelerometers use differential capacitance rather than a single capacitor. This approach significantly improves accuracy, noise immunity, and long-term stability.

In a typical differential design, the proof mass carries arrays of movable comb fingers, while the surrounding frame contains matching fixed fingers. Using many interdigitated fingers in parallel increases the effective capacitance and improves sensitivity. When the device experiences acceleration, the proof mass shifts slightly, increasing the capacitance on one side of the structure while decreasing it on the opposite side. The ASIC measures the difference between these two capacitances, as described by **Equation 1.2**, converting the mechanical displacement into an electrical signal.

Because external influences such as temperature changes, material aging, or supply variations tend to affect both capacitances in a similar way, the differential measurement suppresses many unwanted effects. As a result, the output signal is dominated by true acceleration-induced motion rather than environmental drift.

This differential sensing strategy is a key reason why capacitive MEMS accelerometers achieve high stability, excellent noise performance, and reliable operation across a wide range of conditions.

2.2.5 Air Damping Around the Proof Mass

Although the proof mass moves freely, its motion is not uncontrolled. Gas trapped in the narrow gaps around the MEMS structure creates squeeze-film damping, which resists motion and helps stabilize the response. The detailed role of damping in shaping sensor behavior is discussed later in Section 2.4.

2.2.6 Mechanical Stops: Protecting the Device From Shock

Real-world devices experience sudden impacts such as drops, bumps, collisions, and industrial vibrations. To protect the delicate springs and beams, MEMS accelerometers incorporate mechanical stops near the proof mass. During extreme acceleration, often tens or hundreds of g, the proof mass may momentarily contact these stops. Although this contact can produce a noticeable disturbance in the output signal, it helps prevent structural damage to the sensing elements. As a result, many consumer accelerometers are shock-rated to withstand accelerations of up to $10{,}000g$, despite having proof masses that weigh only micrograms.

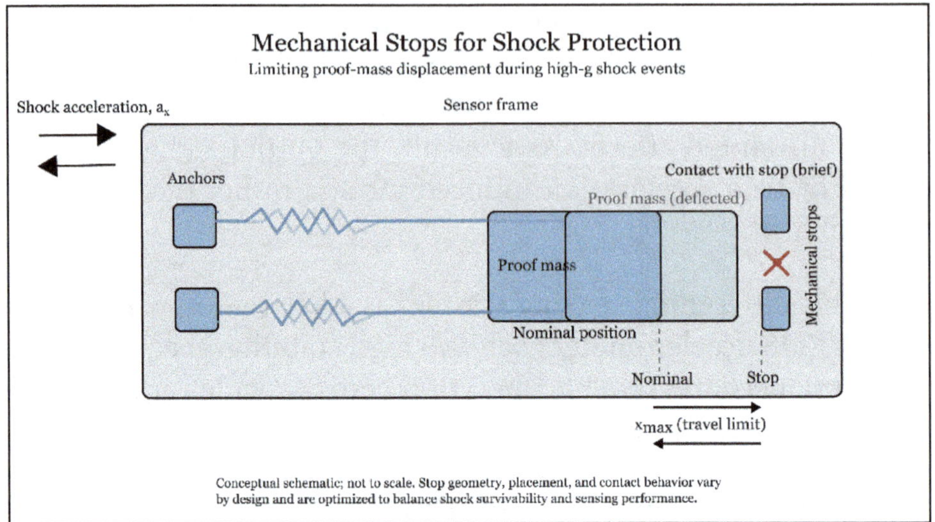

Figure 2.5 Mechanical stops limiting proof-mass displacement during a shock event, showing nominal and deflected positions along the compliant sensing axis.

Figure 2.5 illustrates how mechanical stops protect a MEMS accelerometer during high-g shock events. When a shock acceleration is applied along the compliant sensing axis, the proof mass lags the sensor frame due to inertia and moves relative to its nominal position. If the displacement exceeds the designed travel range, the proof mass briefly contacts the mechanical stops, limiting further motion and preventing damage to the suspension springs and beams. This protective mechanism allows accelerometers with microscopic proof masses to survive extreme shock levels without permanent structural damage.

2.2.7 Multi-Axis Sensing: Measuring X, Y, and Z in a Single Package

Although the sensing structures for the three axes may differ internally, the ASIC keeps each axis electrically independent, allowing the accelerometer to output a_x, a_y, and a_z simultaneously.

This multi-axis capability enables devices to sense orientation, balance, and full three-dimensional motion within a single compact package.

2.3 The Signal Chain: Capacitance to Digital Output

Once the mechanical structure has produced a change in capacitance, the ASIC converts this change into a clean digital signal. This process includes charge-based measurement, analog amplification, filtering, analog-to-digital converter (ADC) conversion, and digital signal processing.

Each stage in the signal chain contributes to accuracy, stability, and usability by reducing noise, compensating for temperature effects, and formatting the output for the host system, such as a microcontroller or application processor.

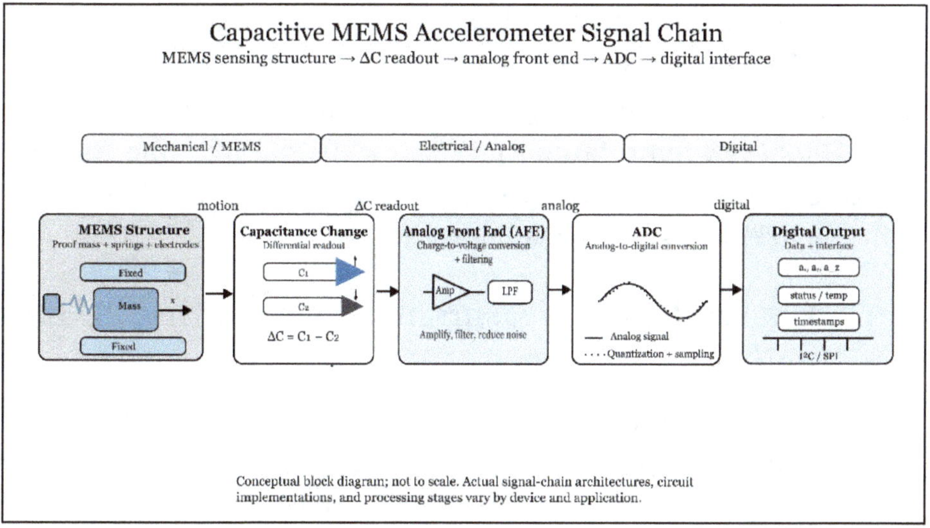

Figure 2.6 Signal chain in a capacitive MEMS accelerometer, showing conversion from differential capacitance (ΔC) to digital output. Schematic illustration only.

Figure 2.6 summarizes the signal chain of a capacitive MEMS accelerometer. Mechanical motion of the proof mass produces a differential change in capacitance, which is converted by the analog front end into a conditioned voltage signal. This signal is then digitized by the ADC and delivered as digital data to the host system, where it can be used for motion detection, orientation, and higher-level processing.

2.3.1 The Capacitive Domain: Sensing the Change

Everything begins with the differential capacitor pair inside the MEMS structure. As illustrated in **Figure 1.4** and expressed in $\boldsymbol{\Delta C}=C$⍰1⍰ $- C_2$, acceleration causes the capacitance C_1 to increase while the capacitance C_2 decreases. The ASIC reads the difference between these two capacitances, $\Delta C = C_1 - C_2$. This difference is the raw signal that encodes the mechanical response of the device.

However, ΔC is extremely small. Typical capacitance values for MEMS comb structures are on the order of $C \approx 1\ to\ 10\ pF$, while the change under $1g$ of acceleration is often only $\Delta C \approx 1\ to\ 100\ fF$. A femtofarad $(fF)\ is\ 1 \times 10^{-15}\ F$, or one quadrillionth of a farad. Detecting such small capacitance changes requires extremely sensitive analog circuitry.

2.3.2 Charge-Based Measurement: Converting ΔC Into Voltage

When acceleration causes the proof mass to move, the resulting displacement produces a differential change in capacitance between opposing capacitor structures. To sense this change, the ASIC applies a known excitation voltage to the capacitors and measures the resulting charge imbalance. Because the stored charge is proportional to capacitance, the differential charge reflects the mechanically induced ΔC.

The analog front end converts this differential capacitance change into a usable voltage signal. In practice, the capacitance change is often normalized to a reference capacitance, resulting in a dimensionless quantity that improves robustness to process and environmental variations. The resulting output can be approximated as:

$$V_{out} \approx \frac{\Delta C}{C_{ref}} \cdot Gain$$ ***Equation 2.1***

In this expression, V_{out} is the output voltage produced by the accelerometer's analog front end, ΔC is the differential capacitance change resulting from proof mass displacement, and C_{ref} is a reference capacitance used to normalize the measurement and reduce sensitivity to absolute capacitance variations. The term $Gain$ represents the overall amplification applied by the analog front end and subsequent signal conditioning stages, setting the scale of the output voltage for a given capacitance change.

This is not a formal equation, but an intuitive way to understand the behavior. The reference capacitor stabilizes the measurement, while the amplifier gain scales the signal into a usable range. The circuitry must be designed to avoid saturating the amplifier while remaining sensitive enough to detect sub-femtofarad changes.

2.3.3 The Analog Front-End (AFE): Amplification and Filtering

The AFE then conditions the tiny voltage signal before it reaches the digital domain. It performs charge amplification, low-pass filtering to remove high-frequency noise, offset correction to compensate for small mechanical asymmetries, temperature compensation for stability, and common-mode rejection. The analog circuitry must be extremely quiet, because even small amounts of electrical noise can obscure the true signal produced by the MEMS structure.

2.3.4 The ADC: Turning Analog Signals Into Numbers

After amplification and filtering, the signal enters an ADC. This is where the accelerometer becomes a digital sensor. Typical ADC resolutions in MEMS accelerometers range from 10-bit and 12-bit to 14-bit, with up to 16-bit resolution in some high-performance devices.

A 12-bit ADC, for example, produces $2^{12} = 4096$ discrete output levels. If the full-scale range is $\pm 2g$, then each least significant bit corresponds to approximately $LSB \approx 4\,g\,/\,4096 \approx 0.001\,g$ per bit. This simple example illustrates how resolution defines the smallest change in acceleration that the sensor can represent digitally.

2.3.5 Digital Signal Processing and Self-Test

After digitization, the accelerometer's ASIC often applies additional digital processing. This can include digital filtering, offset trimming, self-test interpretation, temperature correction, motion detection algorithms, tap and double-tap detection, and step counting or activity classification. These features allow products such as smartphones and wearables to use accelerometers with minimal external processing.

Most accelerometers also include an internal self-test function. In this mode, the ASIC applies a controlled voltage to dedicated self-test electrodes, generating an electrostatic force that deflects the proof mass. The resulting output shift verifies that the mechanical structure and the entire signal chain are functioning correctly. This capability is particularly important in automotive and medical applications.

2.3.6 Output Formatting and Data Delivery

After processing, the ASIC outputs digital acceleration values over a communication interface such as I^2C, SPI, or in some modern sensors, I^3C. Typical outputs include acceleration along the X, Y, and Z axes, along with optional temperature data and status flags.

The devicc may also provide filtered and unfiltered data streams, motion interrupts, freefall detection signals, and FIFO-buffered samples. These features improve system efficiency and reduce the workload on the host processor. The behavior of each stage in this signal chain directly influences the performance characteristics discussed next.

2.4 Internal Dynamics That Shape Performance

Inside every capacitive MEMS accelerometer, the proof mass, springs, electrodes, and on-chip electronics interact in complex ways. These interactions create behaviors that influence how the device responds to acceleration, how stable the output is, and how clean the final signal becomes. Engineers often see these behaviors reflected in datasheet specifications such as noise density, bandwidth, linearity, or resonant frequency, but at their core, they are physical phenomena rooted in the MEMS structure.

2.4.1 Mechanical Damping: Controlling How the Proof Mass Moves

When acceleration acts on the proof mass, it shifts relative to the frame. Because the proof mass is surrounded by air, or sometimes nitrogen, its motion is opposed by damping forces. Most MEMS accelerometers rely on squeeze-film damping, which occurs when gas trapped in the narrow gaps between comb fingers and surrounding structures resists motion. This damping slows the proof mass and helps prevent uncontrolled oscillation.

Damping determines how quickly the proof mass settles after being disturbed. Too little damping allows oscillation or ringing, while too much damping makes the proof mass move sluggishly and reduces responsiveness.

Three regimes of damping are commonly described. In an underdamped system, the mass oscillates after movement; the response is fast but can overshoot (excess displacement). In an overdamped system, there is no oscillation, but the sensor responds slowly. In a critically damped system, the response is the fastest possible without oscillation, producing stable and accurate behavior.

A well-designed MEMS accelerometer aims for behavior close to critical damping. This ensures good stability while keeping the sensor responsive enough to track real-world motion.

2.4.2 Noise: Why Accelerometers Can Never Be Perfectly Quiet

Even when an accelerometer is perfectly still, its output is not completely flat. Instead, it contains small random fluctuations known as noise. This noise arises from thermal motion of the proof mass, electrical noise in the analog circuitry, quantization in the ADC, and environmental influences.

Rather than eliminating noise, which is not feasible, accelerometer design focuses on controlling noise so that it is predictable and well characterized. These noise properties are reflected in datasheet specifications such as noise density.

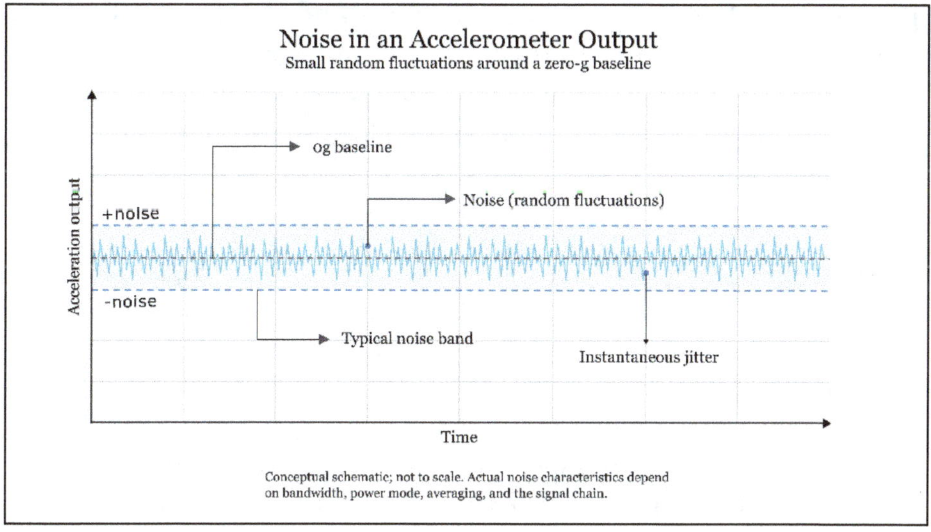

Figure 2.7 Noise in an accelerometer output, showing random fluctuations around a zero-g baseline and the resulting instantaneous jitter.

Figure 2.7 illustrates the presence of noise in an accelerometer output even when no external acceleration is applied. Rather than producing a perfectly flat signal, the output exhibits small random fluctuations around the zero-g baseline due to thermal motion of the proof mass, electronic noise, and quantization effects. Noise describes the overall statistical variation of the signal over time, while instantaneous jitter represents the deviation of a single measurement at a specific moment caused by that noise. These fluctuations define a typical noise band and explain why individual samples vary, motivating the use of filtering, averaging, and bandwidth control in practical systems.

2.4.3 Resonance and Natural Frequency in MEMS Structures

Every spring-mass system has a natural frequency at which it tends to oscillate when disturbed. This frequency is determined primarily by the stiffness of the springs and the size of the proof mass. In a lightly damped MEMS accelerometer, the resonant frequency is very close to the natural frequency, so the two are often treated as nearly the same in practice. A simplified expression for the natural frequency is:

$$f_0 \approx \left(\frac{1}{2\pi}\right) \cdot \sqrt{\frac{k}{m}}$$ ***Equation 2.2***

where k is the spring stiffness (spring constant) and m is the proof mass.

The natural frequency is the frequency at which the proof mass oscillates when it is disturbed and then released, with no continuous external input. The resonant frequency, in contrast, is the frequency at which the system exhibits its largest response when driven by an external periodic excitation. In lightly damped MEMS structures, these two frequencies are very close, so they are often treated as nearly the same in practice.

The natural frequency is built into the structure and cannot be eliminated. The goal of MEMS designers is to ensure that the resonant frequency lies well above the sensor's operational bandwidth, so the proof mass does not oscillate excessively and normal acceleration signals remain accurate and stable.

In practical terms, a low-g accelerometer, such as a $\pm 2\ g$ device, may have a resonant frequency in the 1 to 5 kHz range. A high-g device, designed for hundreds or thousands of g, requires stiffer springs, which raises the resonant frequency significantly.

The resonant frequency becomes important later when we discuss bandwidth in Chapter 3, because usable bandwidth must always remain below the resonant frequency to avoid distortion and instability.

2.4.4 Linearity: A Crucial Feature for Accurate Measurements

Ideally, an accelerometer should produce an output that is directly proportional to the applied acceleration:

$$output = sensitivity \cdot acceleration$$

In practice, real accelerometers deviate slightly from this ideal relationship due to imperfect spring stiffness, non-uniform capacitor structures, small mechanical asymmetries, silicon stress introduced during packaging, and electrical nonlinearities in the analog front-end. These effects produce what engineers call nonlinearity, a measure of how much the output curve deviates from an ideal straight line.

Poor linearity can lead to incorrect interpretation of large accelerations, distorted tilt measurements, and inconsistent sensor response across different measurement ranges. Capacitive MEMS accelerometers generally offer excellent linearity within their specified operating range, especially when compared with older mechanical accelerometer designs.

Figure 2.8 compares an ideal linear accelerometer response with the actual response of a real device. The ideal model assumes perfect proportionality between applied acceleration and output. In practice, mechanical spring behavior, capacitor geometry, and signal-conditioning circuitry introduce small nonlinearities that become more pronounced at higher acceleration levels.

As a result, the actual response curve may deviate above or below the ideal straight line and can even cross it, indicating a change in the sign of the linearity error. This behavior does not imply instability, but rather reflects the limits of linear operation as the sensor approaches its full-scale range.

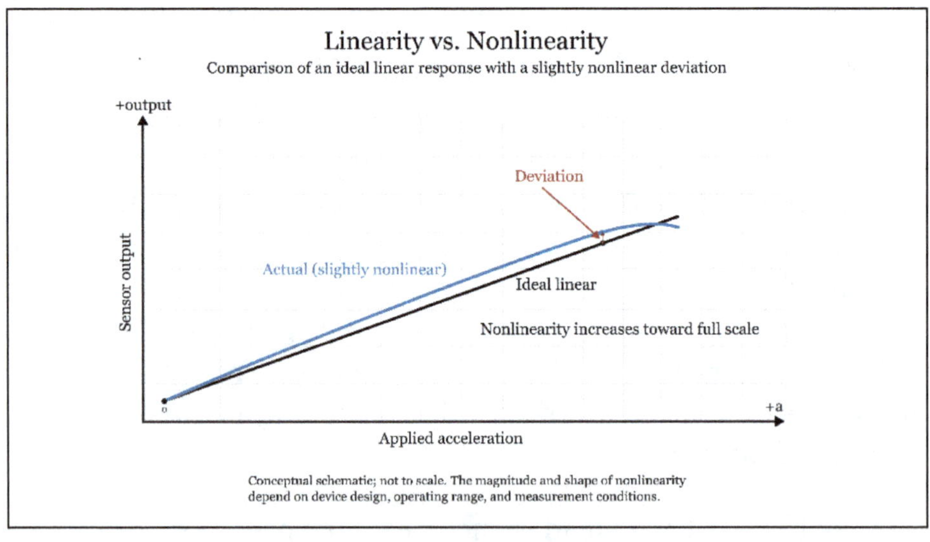

Figure 2.8 Ideal linear response versus actual response showing nonlinearity.

2.4.5 Cross-Axis Behavior: Minimizing Interference

Cross-axis sensitivity refers to the unintended response of an accelerometer axis to acceleration applied along a different axis. In an ideal three-axis accelerometer, each axis measures acceleration independently.

In practice, small mechanical asymmetries, structural misalignment, variations in spring geometry, packaging-induced stress, uneven damping, or electrical coupling can cause acceleration along one axis to produce a minor output on another. High-quality MEMS accelerometers minimize cross-axis sensitivity through careful mechanical layout, symmetric structures, and robust electrical design.

2.4.6 Bandwidth and Settling Behavior

Accelerometers cannot respond to acceleration instantaneously. When acceleration changes, the proof mass requires time to move and settle to a new equilibrium position under a constant (steady) acceleration. For time-varying inputs such as vibration, the proof mass does not settle but continues to move in response to the changing acceleration. This settling behavior is governed by the same mechanical and electrical factors that determine bandwidth, including spring stiffness, damping, resonant frequency, and the analog and digital filtering implemented in the ASIC.

In a mass-spring damper system, bandwidth and settling time are therefore closely related: increasing bandwidth generally reduces both response time and settling time, while limiting bandwidth slows the initial response and increases the time required for the output to settle. This relationship arises because bandwidth is directly related to the system's natural frequency.

A higher natural frequency allows the proof mass to respond more quickly, while a lower natural frequency slows both the initial movement and the return to equilibrium, with the exact settling behavior also influenced by damping.

Response time describes how quickly an accelerometer output begins to change after an acceleration is applied, often characterized by the initial rise of the signal toward its final value. Settling time, in contrast, describes how long it takes for the output to reach and remain within a specified tolerance band around its final value after transients such as overshoot (when the output briefly exceeds its final value) or ringing have decayed.

While both are influenced by bandwidth, damping, and filtering, response time captures the initial speed of reaction, whereas settling time reflects how quickly the measurement becomes stable and reliable.

Ringing refers to the oscillatory behavior that occurs when the proof mass vibrates back and forth after a sudden change in acceleration, rather than settling smoothly to its final position. This behavior is most pronounced when the system is lightly damped. Near resonance, even small excitations can produce large oscillations, leading to distortion and instability in the output.

For this reason, practical accelerometers are designed to operate well below their natural frequency, where the response is more predictable and easier to control. A common design guideline, expressed conceptually as

$$Usable\ Bandwidth\ (BW) < \frac{1}{5} \times f_o$$

which serves as a practical rule of thumb for avoiding resonance, ringing, and instability. For example, if the natural frequency (f_o) is $2\ kHz$, the usable bandwidth may be limited to roughly $300\ to\ 400\ Hz$, providing stable and well-damped behavior suitable for most mobile, wearable, and industrial applications.

2.4.7 Stability Over Time: Drift, Temperature, and Stress Effects

Even a perfectly calibrated accelerometer does not remain unchanged over time. Performance can drift due to internal stress relaxation in the silicon, long-term aging of materials, temperature variation, small shifts in analog circuitry, and packaging stress. These effects are typically small but important because they can gradually shift the sensor's output, alter calibration, and reduce long-term stability.

2.5 Takeaways

In this chapter, we examined how capacitive MEMS accelerometers transform physical motion into electrical and digital signals. We discussed how proof mass motion, differential capacitance, and ASIC signal conditioning work together at microscopic scales.

We began by exploring the mechanical response of the accelerometer under applied acceleration. The proof mass, suspended by compliant springs, moves by extremely small distances, often on the order of nanometers, yet this motion encodes critical information about acceleration. Spring stiffness, damping, and resonant frequency govern how quickly the mass responds and settles, shaping bandwidth and stability.

We then examined the internal sensing architecture. Interdigitated comb structures form differential capacitors whose values change as the proof mass moves. Although these capacitance changes are very small, MEMS fabrication enables them to be highly repeatable. Differential sensing increases sensitivity while rejecting common disturbances such as temperature variation, packaging stress, and long-term drift.

Next, we followed the signal chain from capacitance change to usable output. Charge-based measurement, analog front-end amplification and filtering, ADC conversion, and digital processing each play a role in reducing noise, stabilizing the signal, and delivering acceleration data in a convenient digital form.

Finally, we examined the dynamic behaviors that shape performance, including damping, ringing, resonant frequency, noise, linearity, cross-axis sensitivity, and bandwidth limits. Together, these mechanical and signal-chain concepts form the foundation for understanding how real accelerometers behave and how their performance is specified. In the next chapter, we shift from internal operation to interpreting datasheet specifications and performance metrics.

CHAPTER 3
Key Specifications

Choosing the right accelerometer begins with understanding its specifications. A MEMS accelerometer may appear simple from the outside, but its performance depends on a range of parameters such as measurement range, sensitivity, noise, bandwidth, temperature drift, and power consumption. These specifications are not just numbers on a datasheet. They describe the physical limits of the sensor, how it behaves under real operating conditions, and how well it performs in different applications. By learning how to interpret these parameters, engineers can select accelerometers suited for smartphones, medical devices, automotive systems, industrial equipment, and many other products.

In the previous chapters, we explored what accelerometers measure and how they work internally. In this chapter, we shift our focus from internal mechanisms to measurable performance characteristics. Each specification reflects a different aspect of the accelerometer's behavior, including accuracy, signal quality, response speed, power usage, and performance across temperature and time. Understanding these specifications makes it easier to evaluate accelerometers, compare devices from different manufacturers, and anticipate how a sensor will behave in a real system.

The specifications discussed in this chapter arise directly from the MEMS structures and signal-chain behavior established in Chapter 2 and are treated here strictly as measurable performance parameters rather than structural features.

3.1 Why Specifications Matter

Accelerometer datasheets contain many specifications because each parameter describes a different aspect of measurable performance. Some specifications define how much acceleration the sensor can measure, while others describe noise, response speed, stability, power consumption, or environmental limits. Taken together, these values define the operating envelope of the accelerometer.

Understanding specifications requires recognizing trade-offs. A sensor with lower noise may consume more power. A wider measurement range may reduce sensitivity. Higher bandwidth often increases noise, while low-power modes may increase response time. These trade-offs are inherent to sensor design, not flaws. Learning how to read and balance specifications allows engineers to select accelerometers that behave predictably in real systems.

Figure 3.1 illustrates how accelerometer specifications originate from the physical and electronic behavior of the device. Mechanical motion of the proof mass produces electrical signals that are conditioned, digitized, and processed by the ASIC before being summarized as datasheet specifications. Parameters such as full-scale range, sensitivity, noise density, bandwidth, and power consumption therefore reflect trade-offs that span mechanical design, analog readout, and digital processing under defined test conditions.

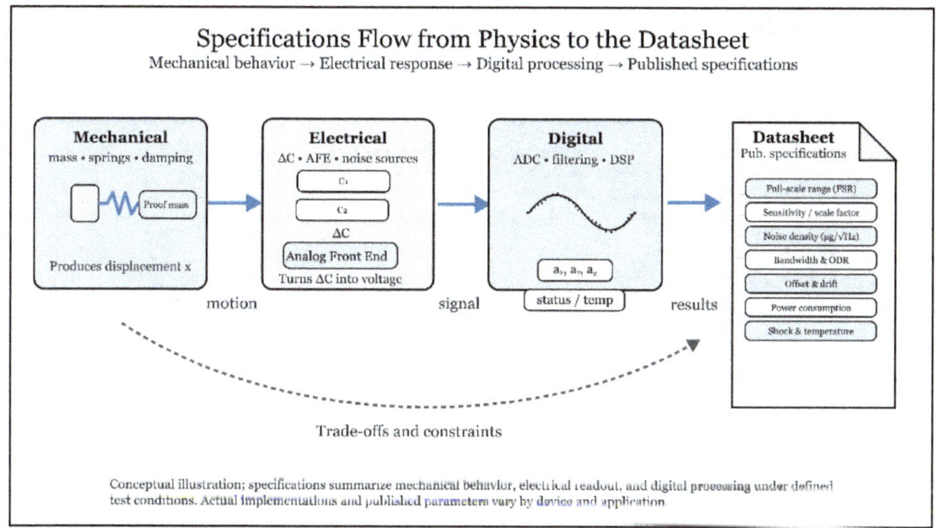

Figure 3.1 Flow of accelerometer specifications from mechanical behavior through electrical and digital processing to published datasheet parameters.

When reading a datasheet, it is important to understand the test conditions under which specifications are measured. Many performance values depend on the selected full-scale range, bandwidth, output data rate (ODR), temperature, supply voltage, or whether filtering is enabled. Two sensors may appear similar on paper, yet behave quite differently in a real system. Paying attention to measurement conditions, and understanding how each parameter influences real behavior, helps avoid common pitfalls in sensor selection and system design.

Engineers often prioritize specifications differently depending on the application. A medical wearable may emphasize low noise and low power consumption. A drone requires fast response and low latency. An industrial monitor may require wide bandwidth and strong vibration handling. An automotive safety device may need robust operation across a wide temperature range and resistance to extreme shock. The same specification can therefore carry different weight depending on system requirements.

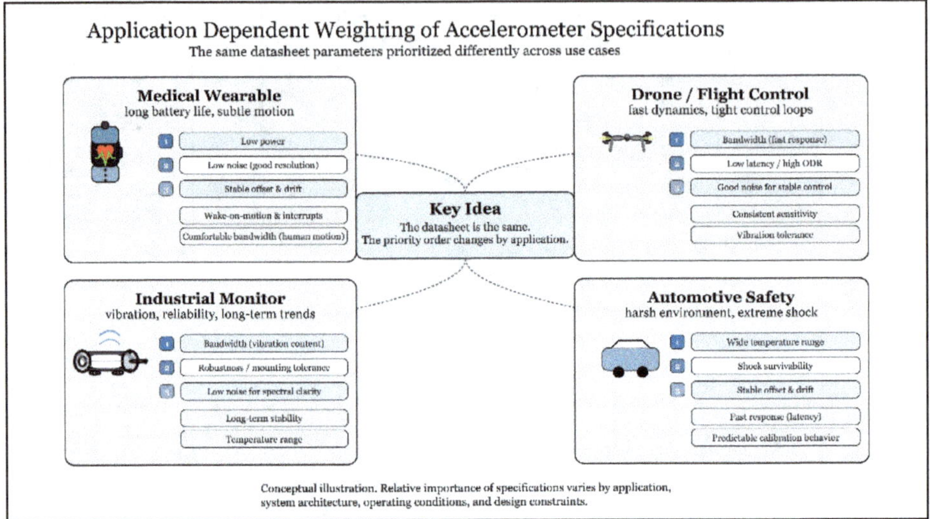

Figure 3.2 Application dependent prioritization of accelerometer specifications, illustrating how the same datasheet parameters carry different weight across use cases.

Figure 3.2 highlights how engineers prioritize accelerometer specifications differently depending on system requirements. While the underlying datasheet parameters remain the same, applications such as medical wearables, drones, industrial monitors, and automotive safety systems emphasize different combinations of power consumption, noise, bandwidth, latency, robustness, and environmental tolerance. Understanding which specifications matter most for a given application helps engineers select sensors that perform predictably in real-world conditions.

Finally, specifications matter because they reflect the real-world behavior of the accelerometer under physical conditions. Understanding each parameter is not merely an academic exercise; it directly influences system design, algorithm development, and overall product reliability.

3.2 Core Measurement Specifications

Capacitive MEMS accelerometers can deliver their measurements as either an analog voltage or a digital value, depending on the sensor design. In both cases, the underlying sensing mechanism is the same: acceleration causes the proof mass to move, the capacitance changes, and the ASIC interprets that change. As a result, many core specifications, such as full-scale range, sensitivity, noise, and bandwidth, apply to both analog and digital accelerometers because they originate from the mechanical and electrical behavior of the underlying structure itself.

Accelerometers with digital output introduce additional specifications, including ODR, digital resolution, and interface characteristics. In this chapter, we focus primarily on capacitive MEMS accelerometers with digital output, which are now the most widely used in modern products.

With this context established, we can now examine the individual specifications that define the capabilities of a capacitive MEMS accelerometer, beginning with the parameters that describe basic measurement behavior.

3.2.1 Full-Scale Range (FSR)

The full-scale range defines the maximum acceleration a sensor can measure reliably before it saturates. For example, a $\pm 2\ g$ accelerometer measures accelerations from $-2\ g$ to $+2\ g$. If the input exceeds this range, the output clips at the limits.

As shown in **Figure 3.3**, different full-scale ranges limit the measurable acceleration span around zero g. Lower full-scale ranges, such as $\pm 2\ g$ or $\pm 4\ g$, provide higher sensitivity and finer effective resolution but saturate more easily when acceleration exceeds the range. Higher ranges, such as $\pm 8\ g$ or $\pm 16\ g$, reduce the likelihood of clipping during large or sudden motions but provide lower sensitivity and coarser effective resolution. Selecting an appropriate full-scale range therefore requires balancing sensitivity and resolution against the expected maximum acceleration in the application.

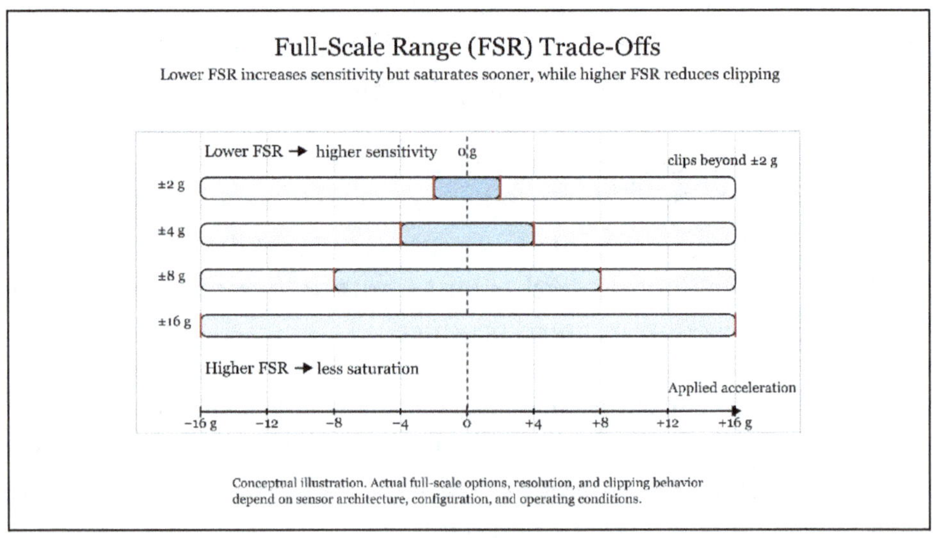

Figure 3.3 Examples of common FSR showing how lower ranges increase sensitivity while higher ranges reduce saturation. Schematic only; actual available full-scale ranges and clipping behavior vary by device and configuration.

For example, choosing a ±16 g range for a wearable device that experiences only ±2 g motion wastes available resolution. Conversely, selecting a ±2 g range for a drone may cause saturation during rapid maneuvers.

Typical full-scale ranges for consumer MEMS accelerometers include $\pm 2\ g, \pm 4\ g, \pm 8\ g$, and $\pm 16\ g$. Engineering and industrial sensors may offer wider ranges such as $\pm 32\ g, \pm 50\ g$, or $\pm 100\ g$ and beyond.

3.2.2 Sensitivity

Sensitivity describes how much the sensor output changes per unit of acceleration. In accelerometers with digital output, sensitivity is usually expressed as:

$$sensitivity \ = \ LSB \ per \ g$$

For example, a sensitivity of $256\,LSB/g$ means the digital output changes by 256 counts for each $1g$ of acceleration.

In accelerometers with digital output, higher sensitivity means more output counts per g, so each digital step corresponds to a smaller change in acceleration. This gives finer resolution and improves the ability to detect small changes in motion. Lower sensitivity means fewer counts per g, so each step represents a larger change in acceleration, reducing resolution but allowing measurement over a wider range.

Consider an accelerometer configured for a $\pm 2\ g$ full-scale range with a 12-bit digital output. The total number of levels is $2^{12} = 4096$, and the full-scale span is $4g$. The ideal sensitivity is therefore:

$$sensitivity \ = \ 4096 \ / \ 4\ g \ = \ 1024\,LSB/g$$

If the same 12-bit ADC is used with a $\pm 16\ g$ range, the full-scale span becomes $32g$, and the sensitivity is:

$$sensitivity = 4096 / 32\,g \approx 128\,LSB/g$$

Thus, in this example, changing the full-scale range from $\pm 2\,g$ to $\pm 16\,g$ alters sensitivity by a factor of eight.

Figure 3.4 illustrates how sensitivity, expressed in counts (LSB) per g, depends on the selected full-scale range. For the same physical acceleration of $1\,g$, a lower full-scale range produces a larger change in digital output, resulting in finer resolution and steeper output slope. A higher full-scale range produces fewer counts per g, which lowers sensitivity and yields coarser resolution, but reduces the risk of saturation during large accelerations. The stair-step behavior represents digital quantization; steps are schematic, and actual devices provide many more discrete output levels.

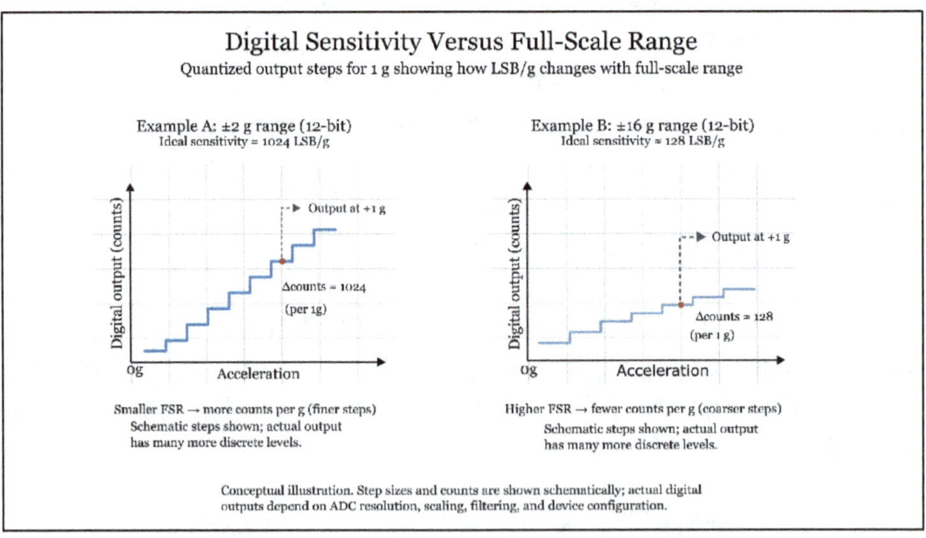

Figure 3.4 Digital output steps corresponding to 1 g for different full-scale ranges, illustrating how sensitivity (LSB/g) changes with FSR.

3.2.3 Resolution

Resolution describes the smallest acceleration change that can be represented by the sensor's digital output and is set by the finite number of available output levels. For example, a 12-bit digital output provides 4096 discrete levels. With a $\pm 2\ g$ full-scale range, the total span is $4g$, corresponding to an ideal resolution of:

$$4\ g\ /\ 4096\ \approx\ 0.001\ g\ per\ count$$

In practice, however, noise determines how much of this theoretical resolution can actually be used. The usable resolution, often referred to as the effective resolution, is typically lower than the theoretical value. Even if the digital step size is small, output noise can be large enough that acceleration changes of only one or two counts cannot be reliably distinguished. Resolution can therefore be improved by selecting a lower full-scale range, increasing digital resolution, or applying filtering to reduce noise.

3.2.4 Noise Density

Noise density represents the random noise in the sensor output, expressed as acceleration noise per square root of bandwidth ($\frac{\mu g}{\sqrt{Hz}}$).

Typical noise density values for MEMS accelerometers vary depending on sensor design, bandwidth, and power mode, and are commonly specified on the order of tens to hundreds of $\frac{\mu g}{\sqrt{Hz}}$.

Figure 3.5 shows a typical noise density curve of a MEMS accelerometer as a function of frequency. At low frequencies, noise density is higher due to $1/f$ noise and decreases as frequency increases, eventually approaching a flat white-noise floor.

The figure also includes an example bandwidth cutoff to illustrate how filtering limits the amount of noise integrated into the final RMS (root mean square) noise value.

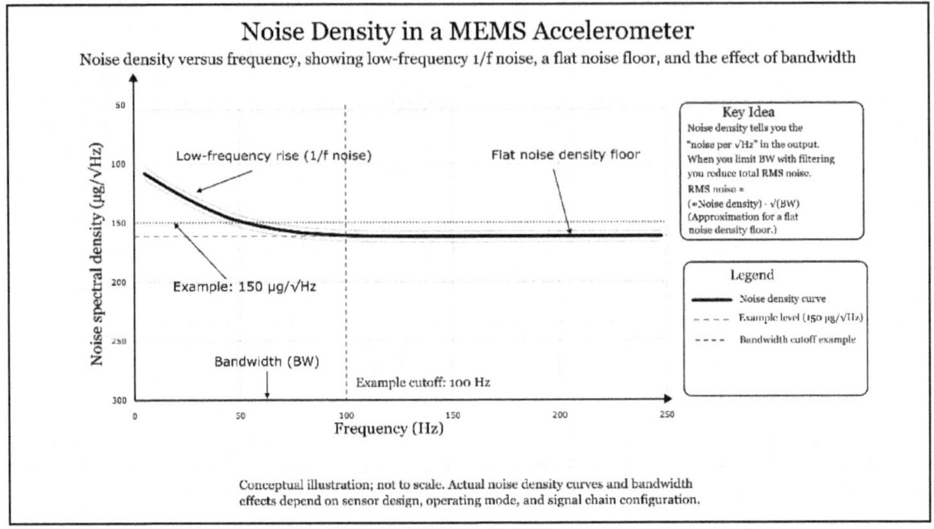

Figure 3.5 Noise density curve of a MEMS accelerometer, showing low-frequency 1/f noise, a flat noise density floor, and the effect of bandwidth on total RMS noise.

This value indicates how noise grows as bandwidth increases. Because the noise contributions are random and partially cancel each other, the total noise increases with the square root of the bandwidth. Noise density determines how much noise remains after filtering. Smaller values correspond to a cleaner signal and better ability to resolve small accelerations.

At low frequencies, noise density increases due to $1/f$ noise, which arises from slow, correlated processes in the sensor and readout electronics, such as charge trapping, bias instability, and drift mechanisms. At higher frequencies, the noise density approaches a flat white-noise floor dominated by uncorrelated thermal and electronic noise sources.

The bandwidth cutoff shown in **Figure 3.5** does not alter the sensor's noise density; instead, it limits how much of this noise is integrated into the final RMS noise, highlighting the role of filtering in practical system design.

3.2.5 RMS Noise (Total Noise Over Bandwidth)

RMS noise represents the total noise present after applying a bandwidth filter.

RMS noise can be estimated using:

$$RMS\ noise \approx noise\_density \cdot sqrt(bandwidth)$$

For example, if the noise density is $150\mu g/\sqrt{Hz}$ and the bandwidth is $100Hz$, then:

$$RMS\ noise = 150 \cdot sqrt(100)$$

$$RMS\ noise = 150 \cdot 10 = 1500\ \mu g = 1.5\ mg$$

This means the filtered output has an RMS noise level of about $1.5\ mg$. RMS noise indicates whether small motions are detectable. If the RMS noise is $1.5mg$, then movements on the order of $0.5mg$ cannot be measured reliably.

Figure 3.6 illustrates how RMS noise appears in the time domain as random jitter around a zero-acceleration baseline after bandwidth filtering. In practice, this filtering may be implemented in the analog domain, the digital domain, or both, depending on the sensor architecture and signal chain. The shaded $\pm 1.5\ mg$ band is a visual reference for the RMS noise level, not a peak-to-peak limit.

Filtering reduces the total noise by limiting the bandwidth over which noise is integrated, but it does not eliminate noise entirely because thermal and electronic noise sources remain within the selected bandwidth. A small example motion of $0.5\ mg$ is shown to emphasize that signals smaller than the RMS noise level are difficult to distinguish reliably from noise.

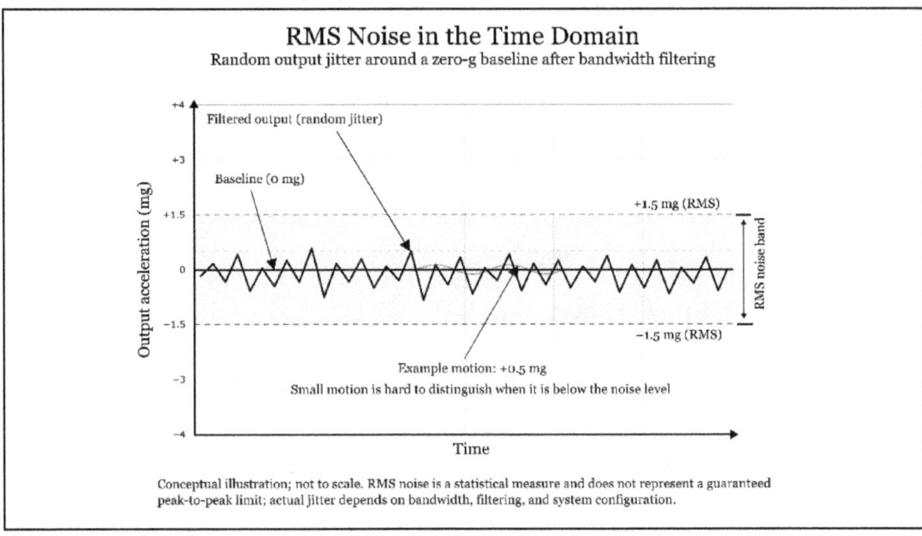

Figure 3.6 RMS noise visualized as random jitter around the baseline after bandwidth filtering.

3.2.6 Bandwidth (BW) and ODR

Bandwidth defines the highest frequency of acceleration that a sensor can measure accurately. Above this frequency, the output becomes attenuated or distorted. Datasheets commonly specify bandwidth values such as $50\ Hz$, $100\ Hz$, $200\ Hz$, or $400\ Hz$ and higher.

The required bandwidth depends on the application, ranging from tens of Hz for human motion to hundreds of Hz or several kHz for machinery vibration and industrial monitoring.

Noise increases as bandwidth increases, so higher bandwidth leads to higher RMS noise. This relationship represents a critical engineering trade-off.

ODR defines how often the accelerometer produces a new measurement. Common ODR settings include $12.5Hz$, $25Hz$, $50Hz$, $100Hz$, $200Hz$, $400Hz$, and $800Hz$ or higher.

3.2.7 Difference between ODR and Bandwidth

Bandwidth refers to the analog front-end limit, while ODR refers to the digital sampling rate. To avoid aliasing, the ODR must be greater than or equal to twice the selected bandwidth, in accordance with the Nyquist condition. If ODR is lower than twice the bandwidth, higher-frequency motion will be under-sampled and may appear in the sampled output as a false lower-frequency signal. In practice, the Nyquist condition can be expressed as

$$ODR \geq 2 \times BW$$

For example, if the bandwidth is set to $100\ Hz$, the ODR must be at least $200\ Hz$. Otherwise, higher-frequency vibrations will fold into lower frequencies and corrupt the measurement.

Figure 3.7 compares proper sampling and undersampling of the same analog input signal. In the left plot, the ODR is high enough relative to the signal frequency, so the sampled points follow the original waveform closely. In the right plot, the ODR is too low, so the sampled points create an apparent lower-frequency waveform that does not represent the true motion. This illustrates how aliasing appears in sampled accelerometer data when the sampling rate is insufficient.

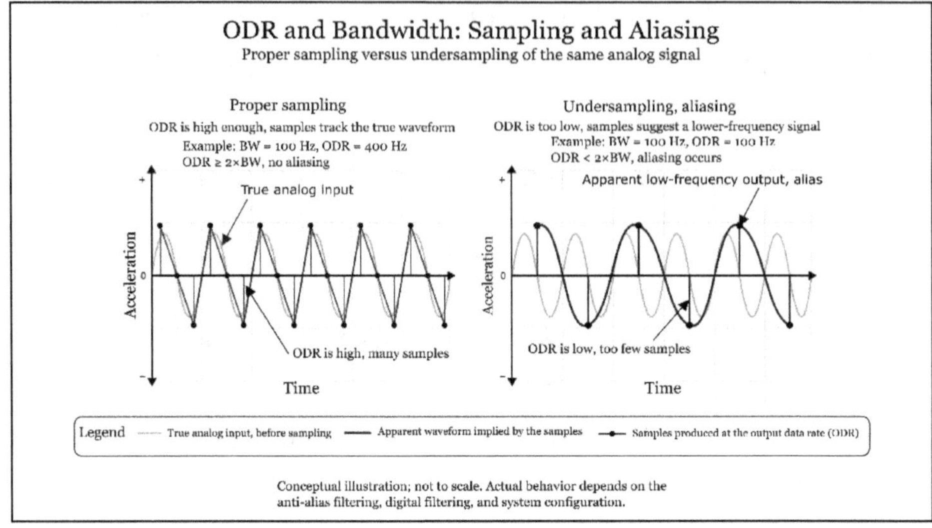

Figure 3.7 Relationship between ODR and bandwidth, illustrating aliasing when the sampling rate is too low.

3.3 Accuracy and Error Sources

The specifications in this section describe the accuracy of the accelerometer, meaning how closely the output matches the true acceleration and how much the reading can drift over time or vary with operating conditions. These parameters often play a more important role in real applications than sensitivity or full-scale range alone, because accuracy determines whether a product behaves consistently across temperature, over time, and under different mounting conditions.

While datasheets provide numerical limits for these error sources, understanding their physical origins makes it easier to interpret the numbers and judge whether a particular accelerometer is suitable for a given application.

3.3.1 Axis Misalignment

In practice, the sensing axes of an accelerometer may not be perfectly orthogonal due to small fabrication tolerances or package alignment errors. This condition, known as axis misalignment, causes acceleration along one axis to partially project onto another axis. Although the effect is usually small, it can introduce measurement errors in applications requiring precise orientation or multi-axis motion analysis.

3.3.2 Non-linearity

Non-linearity describes how much the accelerometer's output deviates from a perfectly linear relationship with applied acceleration. The concept and physical origins of non-linearity were introduced earlier in Section 2.4 and illustrated in **Figure 2.8.**

In the context of accuracy specifications, non-linearity is treated as a bounded error term and is typically specified in datasheets as a percentage of full scale. It represents the maximum deviation of the actual output curve from the ideal straight-line response over the specified measurement range. Datasheets commonly express non-linearity as:

$$non - linearity = (maximum\ deviation\ /\ full - scale\ output) \cdot 100\%$$

Typical MEMS accelerometers specify non-linearity values on the order of about 0.1% to 1% of full scale. If the full-scale range is $\pm 2\ g$ and the maximum deviation from linearity is $0.005\ g$, then:

$$non - linearity = (0.005\ /\ 2) \cdot 100\% = 0.25\%$$

This value indicates the worst-case deviation from an ideal linear response within the specified operating range. Here, the full-scale value refers to the maximum magnitude in one direction (2 g), rather than the total span from $-2\ g$ to $+2\ g$.

3.3.3 Zero-g Offset (Offset Error)

Zero-g offset is the accelerometer's output when no acceleration is applied along a particular axis, other than gravity. For an X axis placed perfectly level, the true acceleration is $0\ g$, yet the output may show a small non-zero value.

This offset arises from small asymmetries in the MEMS structure, residual stress introduced during packaging, slight imperfections in the ASIC, and normal manufacturing tolerances. Offset affects every measurement. A device with a $50\ mg$ offset cannot measure small accelerations accurately unless calibration or compensation is applied.

For example, if an accelerometer reports $+0.025\ g$ on a particular axis when that axis is oriented perpendicular to gravity, the zero-g offset on that axis is $25\ mg$.

3.3.4 Offset Temperature Drift

Offset typically changes with temperature. This behavior is described by the zero-g offset temperature coefficient, usually expressed in $mg/°C$.

Figure 3.8 illustrates two related offset error mechanisms. The left panel shows zero-g offset, where a non-zero output is present even when no acceleration is applied. The right panel shows how this offset changes with temperature, characterized by the zero-g offset temperature coefficient. Together, these effects determine the baseline accuracy of the accelerometer and explain why calibration or temperature compensation is often required in precision applications.

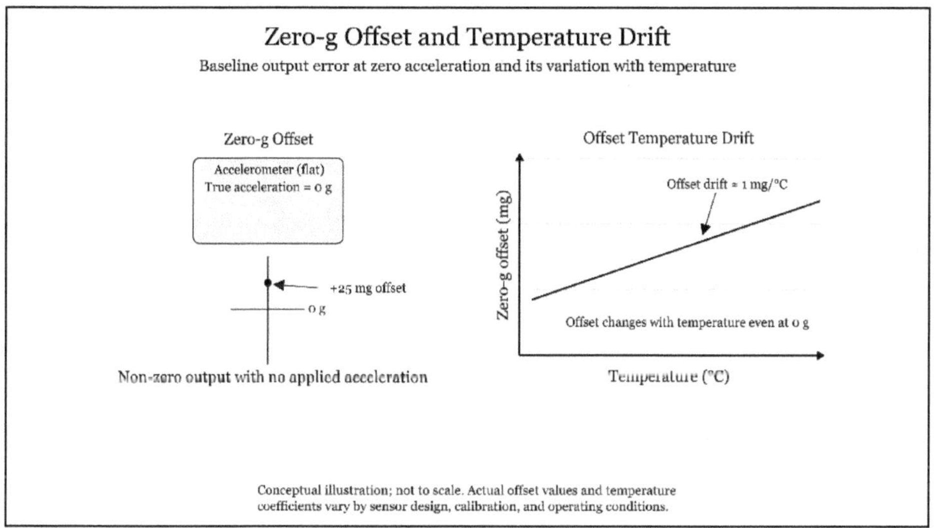

Figure 3.8 Zero-g offset and offset temperature drift in an accelerometer.

For example, if the offset changes by $2\,mg$ for every $1\,°C$ change, the offset drift is $2\,mg/°C$. In automotive, wearable, or outdoor applications, temperature can change by $40\ to\ 80\ °C$ during operation. Even a small offset drift per degree can accumulate into a significant error. If the offset drift is $1\,mg/°C$ and the temperature changes by $30\ °C$, the total drift is:

$$total\ drift\ =\ 1\,mg/°C\ \cdot\ 30\ °C\ =\ 30\,mg$$

This error may be acceptable for some applications but significant for others.

3.3.5 Sensitivity Temperature Drift

Just as offset changes with temperature, sensitivity can also vary with temperature. Datasheets specify sensitivity drift either as a percentage per degree Celsius or as a percentage across the full operating temperature range.

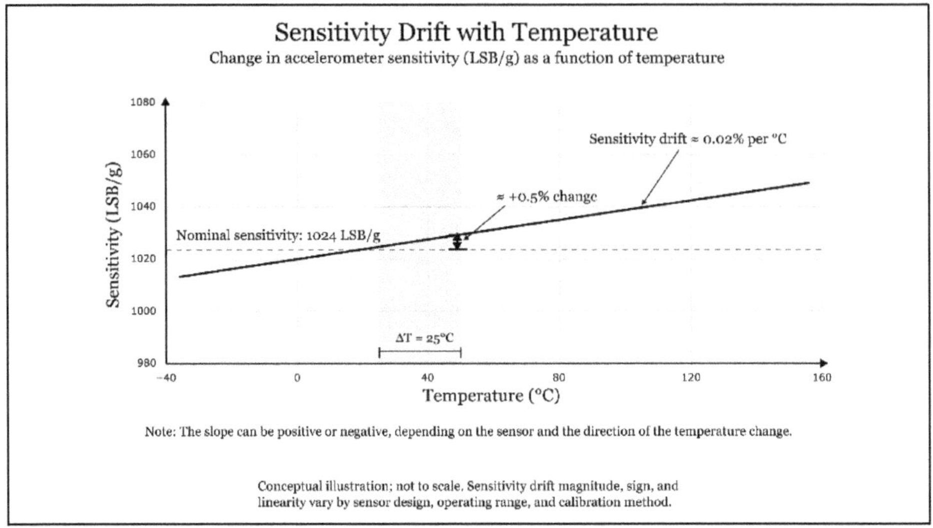

Figure 3.9 Sensitivity drift as a function of temperature in a digital accelerometer.

Figure 3.9 illustrates how an accelerometer's sensitivity, expressed in LSB per g, changes with temperature. The nominal sensitivity is shown as a reference, while the sloped line represents sensitivity temperature drift, expressed as a percentage per degree Celsius. The highlighted temperature interval demonstrates how a modest temperature change can produce a measurable scaling error, emphasizing the need for temperature compensation in precision measurements.

Sensitivity drift arises from changes in spring stiffness with temperature, slight variations in the permittivity of the dielectric, and temperature effects within the analog signal chain.

For example, if the nominal sensitivity is 1024 LSB/g and the sensitivity drift is 0.02% per $°C$, then across a 25 $°C$ temperature change the new sensitivity is calculated as follows. The sensitivity drift is specified as a percentage of the nominal sensitivity, so it must first be applied to the original sensitivity value:

$$0.02\% = 0.0002$$

$$Sensitivity\ change\ per\ °C = 1024 \times 0.0002$$
$$= 0.2048\ LSB/g\ per\ °C$$

$$Total\ sensitivity\ change\ over\ 25\ °C = 0.2048 \times 25$$
$$= 5.12\ LSB/g$$

$$New\ sensitivity = 1024 + 5.12 = 1029.12\ LSB/g$$
$$\approx 1029\ LSB/g$$

This affects scaling accuracy but can be corrected in software when the device provides a temperature output. The sign of the change depends on whether sensitivity increases or decreases with temperature, as well as on the direction of the temperature change.

3.3.6 Cross-Axis Sensitivity

Cross-axis sensitivity describes how acceleration applied along one axis produces an unintended response on another axis. The physical origins of this effect were introduced earlier in Section 2.4 and can be understood by examining the shared proof mass and suspension structures shown in **Figure 2.1**.

In the context of accuracy specifications, cross-axis sensitivity is treated as a bounded error term (a known maximum contribution to measurement error) and is typically expressed as a percentage of the applied acceleration on the intended axis. It reflects imperfect mechanical isolation between sensing directions due to small asymmetries in spring stiffness, electrode geometry, packaging stress, or mounting tilt. Typical datasheet values range from about 1% *to* 5% of full scale, depending on sensor design and packaging.

For example, if 1 g is applied along the Y axis and the X axis reports 0.03 g, the cross-axis sensitivity is 3%.

$$Cross - axis\ sensitivity = \frac{output\ on\ unintended\ axis}{applied\ acceleration} \times 100\%$$

$$Cross - axis\ sensitivity = \frac{0.03\ g}{1\ g} \times 100\% = 3\%$$

3.3.7 Hysteresis

Hysteresis describes the difference in sensor output when acceleration increases compared with when it decreases. For example, if an accelerometer measures +1 g while the acceleration is ramping up, but +0.98 g while ramping down, the 0.02 g difference represents hysteresis.

Hysteresis arises from microfriction within MEMS structures, electrostatic adhesion, mechanical stress relaxation, and imperfections in the analog electronics. In capacitive MEMS accelerometers, hysteresis is typically very small.

Hysteresis means that the sensor output at a given acceleration can depend slightly on whether the input is increasing or decreasing. In other words, the accelerometer does not always follow exactly the same output path in both directions. This matters most in applications where acceleration changes slowly or where the same motion is repeated and precise comparison of the readings is important. Even when the hysteresis value is small, it can contribute to measurement uncertainty when combined with offset, non-linearity, and temperature-related errors.

3.3.8 Bias Instability (Long-Term Stability)

Bias instability refers to low-frequency stochastic variation in the accelerometer offset, causing the zero-g output to wander slowly over time even when the device is stationary. In practice, this behavior appears as a gradual, unpredictable change in the baseline reading over minutes, hours, days, or longer periods, depending on the sensor design and operating conditions.

Typical values for consumer accelerometers range from about $5\ mg$ to $50\ mg$ of drift over time. Industrial or high-end sensors may limit this drift to approximately $1\ mg$ to $5\ mg$.

Bias instability is especially important in low-frequency or quasi-static measurements, where the sensor is expected to hold a stable baseline over long periods. In such cases, a slowly wandering offset can be mistaken for a real change in tilt or acceleration if no correction is applied. This is one reason why long-duration measurements often require averaging, recalibration, or periodic reference updates. Good long-term stability is particularly valuable in precision sensing, condition monitoring, and systems that rely on small changes in acceleration over time.

3.3.9 Total Error Band

Some datasheets combine multiple error sources into a single parameter called total error, total error band, or in-run bias stability. This parameter aggregates contributions from non-linearity, offset, temperature drift, sensitivity drift, hysteresis, and long-term drift.

Figure 3.10 illustrates how multiple accelerometer error sources combine to form a single conservative accuracy bound. Individual contributions such as offset, temperature drift, sensitivity drift, non-linearity, cross-axis sensitivity, hysteresis, and long-term bias instability are stacked around the ideal output to produce a total worst-case error band. This representation reflects how datasheets often specify accuracy using summed limits to guarantee performance under all operating conditions.

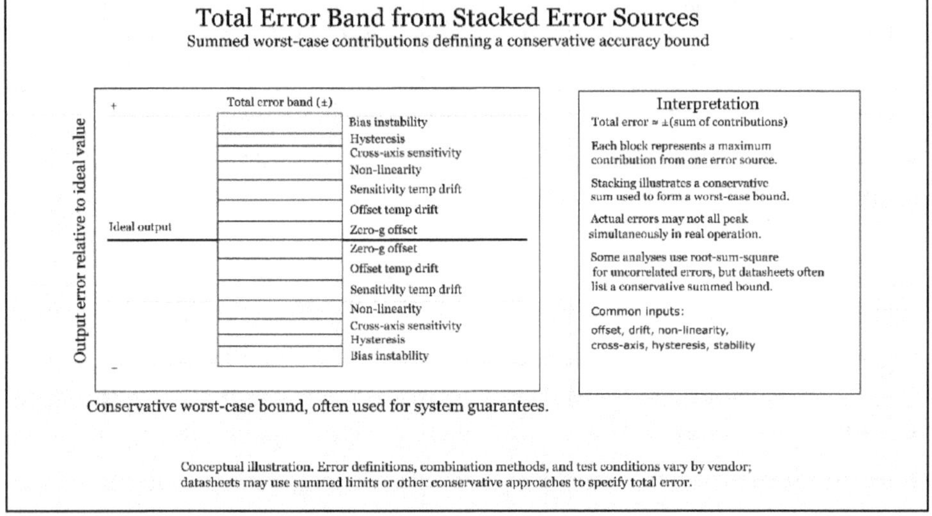

Figure 3.10 Stacked error contributions forming a total worst-case error band.

Total error provides engineers with a single worst-case accuracy bound, often expressed as:

$$total\ error\ =\ \pm(sum\ of\ individual\ contributions)$$

Although conservative, this specification is valuable in systems that must guarantee accuracy under all operating conditions.

In practice, not all error sources reach their worst-case values at the same time or in the same direction. For that reason, the total error band is usually a conservative bound rather than a typical operating error. Even so, it is useful because it helps designers judge whether a sensor can meet system-level accuracy requirements under the full range of expected conditions. When comparing devices, a smaller total error band generally indicates a more predictable sensor, especially in applications that must maintain accuracy across temperature, time, and mounting variation.

3.4 Environmental and Reliability Specifications

Accelerometers are exposed to very different operating environments depending on where they are used. A smartphone experiences gentle everyday motion, while an automotive airbag sensor must survive extreme shocks. An industrial vibration monitor may operate near high-temperature machinery, and a medical wearable must remain reliable despite humidity, body heat, and long periods of continuous use.

Environmental and reliability specifications describe how robust an accelerometer is and whether it can maintain acceptable performance under demanding conditions. These parameters are especially important in safety-critical and long-lifespan applications.

Below are the key environmental specifications that appear in most accelerometer datasheets.

3.4.1 Operating Temperature Range

The operating temperature range defines the temperatures over which the accelerometer is specified to meet the performance and functional requirements listed in its datasheet. Common ranges include $-40\ °C\ to\ +85\ °C$ for consumer and industrial devices, $-40\ °C\ to\ +105\ °C$ for automotive-grade devices, and $-40\,°C\ to\ +125\,°C$ for high-temperature automotive or industrial applications.

Temperature affects MEMS spring stiffness, proof mass movement, capacitance, analog gain, and both offset and sensitivity drift. A device may meet performance specifications at room temperature but drift significantly at higher or lower temperatures, so the specified temperature range must match the expected operating environment.

3.4.2 Shock Resistance

Shock resistance describes how much sudden acceleration the sensor can withstand without permanent damage. Although MEMS structures are delicate, they are well engineered and can survive extreme impacts thanks to mechanical stops and robust silicon anchors.

Typical shock ratings include $1000g, 5000g$, and $10{,}000g$ or more for some designs. These values represent survival limits rather than measurement limits.

Phones, wearables, and handheld devices may be dropped. Industrial sensors may experience impacts during installation or maintenance. Automotive sensors must survive crash events. Even if the accelerometer saturates during an impact, the critical requirement is that it remains structurally intact.

3.4.3 Vibration Resistance

Vibration resistance defines how well an accelerometer operates under continuous or high-frequency vibration. Excessive vibration can excite the resonant frequency, increase noise, stress the MEMS springs, and cause long-term fatigue.

Datasheets typically describe vibration resistance using limits on acceleration level and frequency range. These specifications indicate the vibration conditions the device can tolerate without damage or performance degradation, rather than providing a detailed characterization of vibration behavior.

In industrial or automotive environments, continuous vibration is common. Drone motors, vehicle engine blocks, and production machinery can expose sensors to strong vibration for extended periods. Good vibration resistance ensures the accelerometer maintains accuracy and avoids long-term structural fatigue.

3.4.4 Humidity and Environmental Scaling

Although MEMS structures are protected inside the package, humidity and condensation can still influence internal stress and long-term reliability. Some sensor packages incorporate moisture sealing, protective coatings, or hermetic ceramic enclosures, which are typically used in aerospace applications or medical implants.

Most consumer-grade accelerometers are designed to operate in typical indoor or outdoor humidity conditions but are not intended for immersion or extreme moisture exposure unless additional housing protection is provided.

Moisture can introduce package stress, accelerate long-term drift, and promote corrosion of bond pads or interconnects if the device is not adequately sealed. Many accelerometers are specified for operation up to relative humidity levels of about $85\%\ to\ 95\%$ non-condensing, typically at temperatures around $40\ °C\ to\ 60\ °C$. Exposure to condensation, immersion, or sustained high humidity beyond these limits can degrade performance over time or lead to permanent failure unless additional protective coatings or sealed system-level enclosures are used.

3.4.5 Mechanical Stress Sensitivity

Mechanical stress sensitivity describes how much the accelerometer output changes when mechanical stress is applied to the package or the PCB. Stress can arise from mounting strain, PCB bending, solder reflow, enclosure pressure, or thermal expansion mismatch. Even small amounts of stress can shift the zero-g offset or affect sensitivity. Mounting a sensor too close to a screw hole can cause the PCB to flex, resulting in a measurable change in output.

In precision applications, sensors should be mounted on rigid areas of the PCB, away from flex points, and using the recommended footprint and soldering patterns. Mechanical stress sensitivity is rarely emphasized in consumer datasheets but is critically important in professional designs.

3.4.6 ESD Protection

Electrostatic discharge (ESD) ratings indicate how well an accelerometer can withstand electrical shocks that occur during manufacturing, handling, or assembly. Typical ratings follow established models such as the Human Body Model (HBM) and the Charged Device Model (CDM). Common values range from about $\pm 2\ kV$ to $\pm 4kV$ for HBM and from $\pm 500V\ to\ \pm 1000V$ for CDM.

ESD damage does not always cause immediate failure but can degrade analog performance, increase noise, shift offsets, or shorten device lifespan. Adequate ESD protection helps ensure robust handling, reliable assembly, and long-term operation.

3.5 Electrical and Operational Specifications

Beyond measurement accuracy, accelerometers must operate within the constraints of the system they are part of. Power consumption, supply voltage, internal operating modes, startup time, and built-in self-test capabilities can be just as important as sensitivity or noise, especially in battery-powered devices, medical sensors, or automotive systems where reliability and timing are critical.

The specifications in this section describe how the accelerometer behaves electrically and how it interacts with the host system. Understanding these parameters helps engineers integrate accelerometers efficiently and select the most appropriate device for a given application.

3.5.1 Current Consumption

Current consumption describes how much current the accelerometer draws in different operating modes. Digital MEMS accelerometers typically specify current for active mode, low-power mode, ultra-low-power or wake-on-motion mode, and standby or power-down mode. Typical values may include about 5 μA to 20 μA in active mode, 1 μA to 5 μA in low-power mode, below 1 μA in wake-on-motion operation, and nanoamp (nA)-level current in standby.

Battery-powered systems often allocate only a few microamps (μA) for continuous sensing. Lower current consumption extends battery life and reduces the load on the main processor, especially when the accelerometer can remain in a low-power monitoring state and wake the host system only when motion is detected.

3.5.2 Power Modes and Wake-on-Motion

Most accelerometers include multiple power modes to balance performance with energy use.

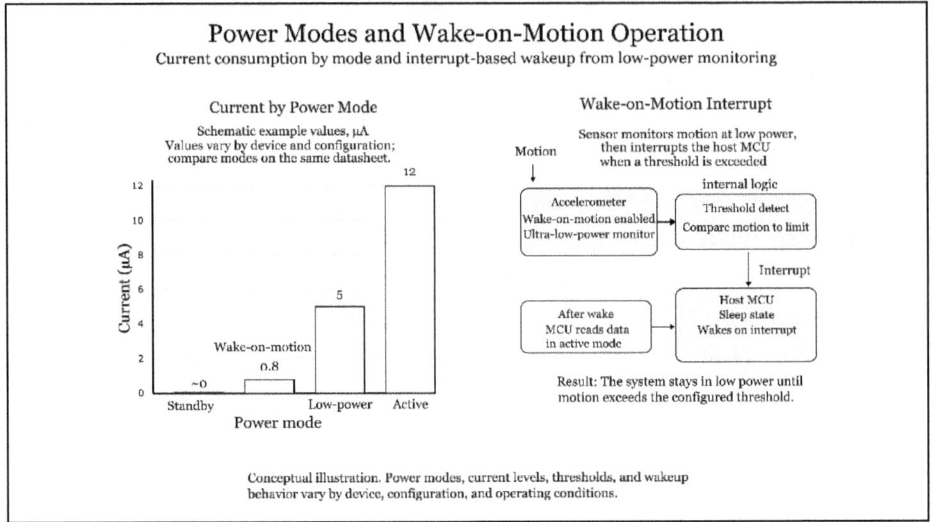

Figure 3.11 Current by power mode and wake-on-motion interrupt behavior in a typical digital accelerometer.

Figure 3.11 summarizes two related aspects of low-power accelerometer operation. The left panel compares typical current consumption across standby, wake-on-motion, low-power, and active modes, highlighting the large reduction in current available in motion monitoring modes. The right panel shows the wake-on-motion workflow, where the accelerometer monitors motion using low power internal logic and asserts an interrupt when a threshold is exceeded, allowing the host microcontroller to remain asleep until needed. This combination enables continuous motion awareness while minimizing overall system power consumption.

These modes adjust parameters such as ODR, bandwidth, internal filtering, analog front-end behavior, and digital processing activity. In wake-on-motion, sometimes referred to as a wake-up mode, the accelerometer operates at extremely low power and monitors motion using simplified internal logic. When motion exceeds a user-defined threshold, the device generates an interrupt to wake the main microcontroller. Wake-on-motion significantly reduces overall system power consumption, making continuous motion monitoring practical for devices powered by small batteries.

A wearable device such as a smartwatch or fitness tracker may operate the accelerometer at approximately 0.8 μA in wake-on-motion mode and 12 μA in active mode. The sensor remains at the lower current level until the wearer begins to move.

3.5.3 Self-Test (Built-In Test Capability)

Digital MEMS accelerometers often include a self-test function that applies an internal electrostatic force to the proof mass. This force simulates a known acceleration and allows the sensor's response to be verified without external motion.

Figure 3.12 illustrates the principle of accelerometer self-test operation. During normal operation, no internal force is applied and the proof mass remains centered, resulting in balanced differential capacitance. When self-test is enabled, an internal electrostatic force deflects the proof mass by a known amount, producing a predictable output shift specified in the datasheet.

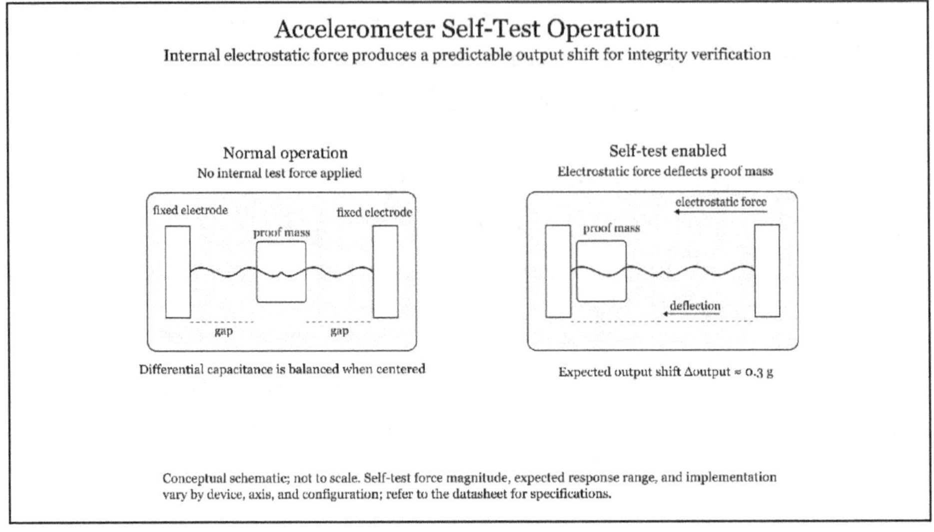

Figure 3.12 Accelerometer self-test applying an internal electrostatic force to the proof mass.

During normal operation, no internal force is applied and the proof mass remains centered, producing a balanced differential capacitance. When the self-test function is enabled, an internal electrostatic force deflects the proof mass by a known amount. This controlled deflection produces a predictable shift in the output signal.

Comparing the measured response with the expected behavior helps verify the basic functionality of the MEMS mechanical structure, analog front end, and digital signal chain. Self-test is therefore useful for periodic integrity checks and can help detect damage or failure in field deployments.

3.5.4 Self-Test Response Interpretation and Tolerance

Self-test operates by applying a known internal electrostatic force to the proof mass, producing a predictable output shift that is specified in the datasheet. The magnitude of this response is defined as a range rather than a single value to account for normal variation among MEMS structures, analog circuitry, and operating conditions.

Because self-test temporarily alters the accelerometer output, it must be performed under controlled conditions. External motion, vibration, or mechanical stress can distort the measurement and lead to incorrect conclusions. Self-test should therefore be executed only when the device is stationary and after the signal chain has fully settled.

Datasheet self-test specifications are typically provided as minimum and maximum acceptable output shifts. These limits represent verified nominal bounds across production devices, temperature, and configuration. If the measured response during self-test falls anywhere within this specified range, the accelerometer is considered to be functioning correctly.

Self-test verifies basic sensor integrity but does not validate offset accuracy, sensitivity calibration, axis alignment, noise performance, or long-term stability. These characteristics must be addressed separately through calibration and system-level validation.

Figure 3.13 illustrates the concept of a valid self-test response range with tolerance limits. Responses near the minimum or maximum limits are still valid and do not indicate failure when they remain within the specified bounds. Interpreting self-test as a strict pass/fail check requires comparing the measured result against these datasheet limits.

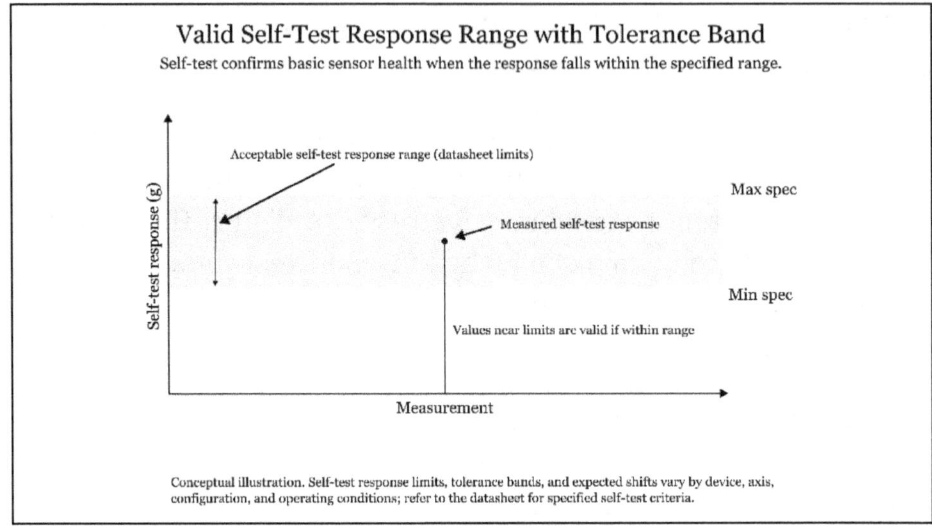

Figure 3.13 Valid self-test response range with tolerance band.

3.5.5 Startup Time (Turn-On Time)

Startup time describes how long the accelerometer requires after power-up before its output becomes valid. Typical values range from about 1 ms to 10 ms for consumer devices and from 10 ms to 30 ms for some automotive or industrial designs.

In duty-cycled systems, long startup times reduce power efficiency because the device must remain powered longer before valid measurements are available. In safety systems, slow startup can delay critical detection, while in motion-triggered applications startup latency affects responsiveness. Accelerometers with shorter startup times are therefore easier to use in low-power systems where the device is frequently powered down and re-enabled.

3.5.6 Supply Voltage and I/O Voltage

Accelerometers specify both the supply voltage (V_{dd}), which powers the MEMS structure and ASIC, and the I/O voltage

(V_{ddIO}), which defines logic-level compatibility with the host system. Typical values for V_{dd} range from about $1.6\,V\ to\ 3.6\,V$, while V_{ddIo} commonly supports levels such as $1.2\,V, 1.8\,V, 2.8\,V, or\ 3.3\,V$. Some sensors allow V_{ddIo} to be supplied independently, enabling compatibility with low power microcontrollers or external voltage-level translation.

The supply and I/O voltages set the allowable operating range of the accelerometer, influence noise performance and power consumption, and ensure reliable digital communication with the host MCU. Operating at lower voltages generally reduces current consumption, which is especially beneficial for battery-powered devices.

3.5.7 Output Format (Digital Resolution and Data Representation)

Accelerometers that provide digital data output report signed numerical values that correspond to measured acceleration. Important aspects of the output format include resolution, such as 10-bit, 12-bit, or 16-bit depth, the LSB-per-g scaling factor, two's complement representation, scaling that depends on the selected full-scale range, and whether the data is provided as raw or filtered output.

It is important to distinguish between the digital word length used to transmit data and the effective measurement resolution of the sensor. Many accelerometers provide output data in 16-bit registers even though the actual usable resolution may be lower, depending on the sensor architecture, noise floor, internal filtering, and signal-processing chain.

Resolution and data format determine the smallest detectable change in acceleration, the usable dynamic range, how easily raw data can be converted into physical units, and compatibility with host-side algorithms. Higher effective resolution allows smaller acceleration changes to be represented in the output data.

Higher digital resolution increases the number of discrete output levels available to represent acceleration, reducing the acceleration change associated with each count. As resolution increases, the smallest detectable change in acceleration becomes smaller, improving measurement granularity, particularly at lower full-scale ranges. This finer quantization allows more precise digital processing and improves compatibility with algorithms that rely on small signal variations, without changing the underlying sensor physics.

3.5.8 Filtering Options (Analog and Digital)

Digital accelerometers typically include an analog low-pass filter before the ADC, followed by digital filtering after conversion. The analog filter helps limit high-frequency content before sampling, while digital low-pass and digital high-pass filters shape the signal further in the digital domain. Selectable bandwidth settings are usually implemented through these analog and digital filter choices, allowing the effective bandwidth to be tailored to the application.

These filters reduce noise, help prevent aliasing, and allow performance to be adjusted for different use cases. For tilt measurement, using a low bandwidth results in lower noise and a smoother output. For vibration monitoring, a higher bandwidth is required to capture high-frequency motion. Engineers must select filtering settings that balance noise performance, responsiveness, and power consumption for the intended application.

3.5.9 FIFO and Data Management

Many modern accelerometers include a FIFO (First-In, First-Out) buffer that stores samples internally. Typical FIFO sizes range from 32 samples to 128 samples, with some devices supporting 512 samples or more.

FIFO buffering reduces interrupt load on the microcontroller, enables burst reads, supports precise sample timing, and improves overall efficiency in low power systems.

3.6 How to Evaluate an Accelerometer Using Its Specifications

Now that we have explored the key specifications of capacitive MEMS accelerometers, including measurement range, sensitivity, noise, bandwidth, drift, power consumption, and environmental limits, we can bring everything together into a practical engineering framework. Evaluating an accelerometer requires understanding not only each specification individually, but also how the specifications interact and how they influence performance in a real system.

Selecting the “best” accelerometer depends entirely on the application. A device well suited for a wearable may be unsuitable for an industrial vibration monitor, and a sensor optimized for automotive safety systems may be excessive for a smartphone. This section provides a structured approach to interpreting specifications so you can select the most appropriate accelerometer for your design.

3.6.1 Identify the Required Full-Scale Range (FSR)

When evaluating an accelerometer, the selected full-scale range should comfortably encompass the maximum expected acceleration in the target application, including transients and shock events. The chosen range must be large enough to avoid saturation during normal operation, yet small enough to preserve adequate sensitivity and resolution.

Rather than focusing on how full-scale range affects sensitivity or resolution, which were discussed earlier in this chapter, the key evaluation question is whether the selected range provides sufficient margin for real-world conditions. For example, wearable and motion tracking applications typically operate within $\pm 2g$ to $\pm 8g$, while industrial vibration monitoring or automotive crash detection may require ranges of $\pm 16g$ or higher.

An appropriate full-scale range balances robustness against unexpected acceleration peaks with the need for usable signal resolution. Selecting the smallest range that reliably captures all expected motion, including rare but realistic extremes, generally results in the most effective overall system performance.

3.6.2 Evaluate Noise and Resolution Together

Noise and resolution must be considered together when determining whether an accelerometer can detect the motion of interest. High digital resolution alone is not sufficient if noise obscures small signals, and low noise provides limited benefit if resolution is too coarse to represent it meaningfully.

If the smallest acceleration you need to detect is $a_{\min} = 0.5mg$, RMS noise should be well below $0.5mg$ to ensure reliable detection. In applications such as vibration monitoring, bandwidth and frequency response may be more important than achieving the lowest possible noise floor.

For slow or quasistatic measurements such as tilt sensing, RMS noise below about $1\ mg$ is desirable. For general mobile and consumer applications, RMS noise in the range of $5\ mg\ to\ 10\ mg$ is often acceptable.

3.6.3 Check Bandwidth and ODR

In some accelerometers, bandwidth is fixed as a fraction of the ODR, so selecting the ODR implicitly determines the usable bandwidth. In other designs, bandwidth and ODR can be configured independently using internal digital filtering and decimation. In all cases, the selected bandwidth and ODR must remain consistent with the Nyquist condition discussed in Chapter 3.

3.6.4 Match Bandwidth to the Application

Different applications require different bandwidths. Human motion and tilt sensing typically fall below $20\ Hz$. Consumer gesture detection often uses bandwidths between $50\ Hz\ and\ 100\ Hz$. Robotics and control applications commonly require $100\ Hz\ to\ 200\ Hz$, while drone stabilization and high-dynamic systems may use $200\ Hz\ to\ 400Hz$ or higher. Industrial vibration monitoring can extend into hundreds of Hz or beyond, depending on the machinery and fault modes of interest.

Select the lowest bandwidth that fully captures the motion of interest while maintaining adequate margin between bandwidth and ODR. This minimizes noise, reduces processing load, and avoids aliasing without sacrificing responsiveness.

3.6.5 Analyze Temperature Drift and Stability

Temperature-related drift should be evaluated in terms of its impact on overall measurement stability across the expected operating range. Rather than focusing on the underlying drift mechanisms, the goal at this stage is to determine whether temperature-induced changes will compromise system accuracy under real-world conditions.

This consideration is particularly important for applications exposed to wide or uncontrolled temperature variations, such as automotive systems, outdoor equipment, wearable medical devices, and industrial machinery near heat sources.

Engineers must therefore estimate how much the sensor output may change over the anticipated temperature swing, for example from $40\ °C\ to\ 60\ °C$, and determine whether that variation remains within the allowable error limits of the application.

If the expected drift exceeds acceptable limits, mitigation strategies may include device-level calibration, software-based temperature compensation, or selecting an accelerometer with tighter temperature specifications.

3.6.6 Consider Power Consumption and Power Modes

Power consumption should be evaluated in the context of the system's operating pattern rather than in isolation. Battery-powered products such as wearables and IoT devices often rely on low-power monitoring or wake-on-motion modes to minimize average current, while robotics and industrial systems may prioritize continuous high sampling-rate operation over power savings.

Select an accelerometer whose power and performance trade-offs are compatible with the system's duty cycle, including how often the sensor is active, how frequently data is processed, and how long the system is expected to operate between power cycles.

3.6.7 Check Non-Linearity and Cross-Axis Sensitivity

Non-linearity and cross-axis sensitivity should be reviewed in terms of their impact on multi-axis accuracy rather than as isolated specifications. These errors can introduce systematic deviations in tilt, orientation, and posture measurements, especially when multiple axes are combined.

For applications requiring accurate tilt or orientation estimation, non-linearity below about 0.5% and cross-axis sensitivity below about 2% are generally desirable. In applications where absolute accuracy is less critical, higher values may be acceptable if other system requirements are met.

3.6.8 Verify Shock, Vibration, and Environmental Limits

Shock, vibration, and environmental limits should be reviewed as part of overall system robustness, even when extreme conditions are not expected during normal operation. Devices may still experience unexpected mechanical stress during handling, installation, startup, or fault events. Examples include drone landings, accidental drops of handheld devices, machinery startup vibrations, and automotive crash pulses.

Select an accelerometer with shock, vibration, and environmental ratings that exceed the worst credible conditions the system could encounter, not just its intended operating environment.

3.6.9 Look at System-Level Factors

Beyond the sensor specifications, system-level considerations often determine real-world performance. These include the host microcontroller's processing capability, the use of interrupts versus polling, FIFO buffering requirements, proximity to electrically noisy components, potential PCB flexing that may introduce mechanical stress, and the sensor's physical mounting relative to the axes of interest.

In practice, a midrange accelerometer may outperform a higher-end device if it integrates more smoothly into the system or behaves more predictably under actual operating conditions.

3.6.10 Compare Sensors Using a Weighted Criteria Approach

Engineers often compare accelerometers by assigning weights to a small set of decision criteria based on the application's priorities. Typical criteria include noise performance, power consumption, temperature stability, bandwidth and latency, cost, shock and vibration tolerance, package constraints, and interface compatibility such as I^2C or SPI.

Figure 3.14 illustrates a criteria-based comparison of accelerometers across multiple performance dimensions within a multi-criteria evaluation framework. Each axis represents a normalized performance score derived from datasheet specifications, allowing different criteria to be compared on a common scale. Each axis represents a normalized performance score for a specific criterion, such as noise performance, power consumption, temperature stability, bandwidth and latency, shock and vibration tolerance, size and integration, interface and features, and cost and availability. Larger enclosed areas indicate a better overall fit for the selected priorities. The radar chart does not apply weights directly, but provides a visual summary that helps engineers identify tradeoffs and support structured sensor selection decisions.

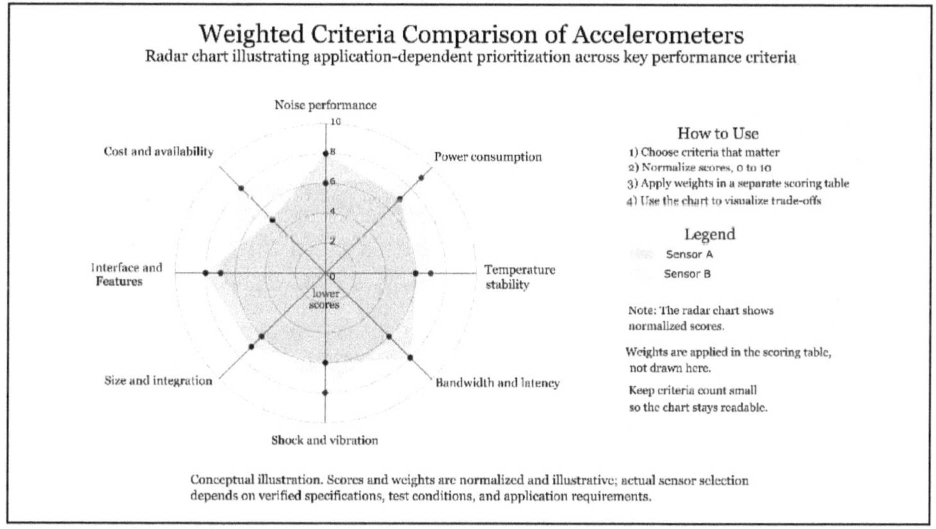

Figure 3.14 Multi-criteria comparison of two accelerometers using a radar chart.

The weighting depends on the use case. Wearables typically emphasize power and noise, drones emphasize bandwidth and latency, industrial systems emphasize stability and robustness, and automotive designs often emphasize temperature behavior, shock survival, and reliable self-test. A simple multi-criteria comparison helps make these tradeoffs visible and supports identification of the most suitable sensor.

3.6.11 Always Consider Calibration Needs

Calibration requirements should be evaluated early, as some accelerometers rely on user calibration for offset, sensitivity, cross-axis effects, or temperature behavior, while others include factory calibration. The level of calibration support can significantly influence system complexity and long-term accuracy.

In high-accuracy applications, effective calibration and compensation strategies often have a greater impact on performance than small differences in nominal datasheet specifications.

3.7 Takeaways

In this chapter, we explored the key specifications that define the performance of a capacitive MEMS accelerometer. Unlike the internal mechanics discussed in Chapter 2, specifications describe measurable characteristics, numerical values that indicate how the sensor behaves in real systems. These parameters help engineers determine whether an accelerometer is suitable for a given application, compare devices from different vendors, and anticipate performance under various operating conditions.

We examined why specifications matter and how they capture different aspects of sensor performance, including measurement range, sensitivity to small motions, signal noise, accuracy across temperature, power consumption, and robustness against shock, vibration, and environmental stress. Understanding these parameters allows engineers to interpret sensor behavior rather than relying solely on headline numbers.

We then reviewed the core measurement specifications that describe how accelerometers quantify acceleration, including full-scale range, sensitivity, resolution, noise density, RMS noise, bandwidth, and ODR. These parameters determine how well the sensor captures real motion, detects small acceleration changes, avoids saturation, and delivers usable data.

We also examined accuracy-related specifications such as non-linearity, cross-axis sensitivity, zero-g offset, temperature drift, and long-term stability, which explain why real-world outputs deviate from ideal behavior. In addition, we discussed electrical and operational parameters that affect system integration, including power consumption, operating modes, wake-on-motion features, startup time, filtering options, digital resolution, and self-test capabilities.

Finally, we reviewed environmental and reliability specifications, including operating temperature range, shock resistance, vibration tolerance, humidity protection, mechanical stress sensitivity, and ESD robustness, which determine whether the sensor can survive the conditions of its intended application.

By combining these specifications, engineers can evaluate accelerometers using a structured approach. Selecting a sensor is not simply about choosing the lowest noise or the widest range, but about balancing trade-offs and matching specifications to application requirements. Wearables, drones, medical devices, automotive systems, and industrial equipment each impose different priorities, and the most appropriate accelerometer is the one whose characteristics align with the system's goals.

With this understanding of accelerometer specifications and how they relate to real sensor behavior, you now have the foundation needed to interpret accelerometer datasheets effectively. In the next chapter, we focus on how these specifications are presented in datasheets, including how to read tables, curves, test conditions, and footnotes, and how to extract the information that matters most for evaluating and selecting sensors.

CHAPTER 4
How to Read a Datasheet

Accelerometer datasheets can appear dense at first glance. They contain numerous specifications, electrical tables, performance curves, timing diagrams, and configuration details that must be interpreted correctly for a sensor to perform reliably in a real system. While earlier chapters explained what accelerometers measure, how they work, and how specifications shape performance, this chapter focuses on how to read a datasheet with confidence and extract the information that truly matters for your design. Understanding a datasheet is an essential engineering skill, whether you are selecting a sensor, comparing vendors, or verifying that a device meets application requirements.

A datasheet is more than a collection of numbers; it is a compact summary of an accelerometer's capabilities, limits, and test conditions. When interpreted correctly, it allows engineers to anticipate real-world behavior long before writing code or building hardware.

Accelerometer datasheets describe performance limits and operating behavior that originate from the MEMS structures and signal chain described in Chapter 2. This chapter does not re-explain those internal mechanisms. Instead, it focuses on how manufacturers express MEMS behavior through specifications, tables, curves, test conditions, and footnotes, and how engineers should interpret that information when evaluating and integrating a device.

4.1 Understanding the Structure of a Datasheet

Before examining individual specifications, it is essential to understand the overall structure of a typical accelerometer datasheet. Although manufacturers present information in slightly different formats, most datasheets follow a common pattern because they must convey the same fundamental details, including electrical behavior, performance characteristics, mechanical layout, timing information, configuration instructions, and packaging details. Once the purpose of each section is clear, interpreting a datasheet becomes far more intuitive.

A datasheet is essentially a compact engineering contract between the manufacturer and the designer. It defines what the sensor can do, how it behaves under specific conditions, which limits must not be exceeded, and how it should be integrated into a system. None of this information exists in isolation. Electrical limits affect noise performance, bandwidth influences resolution, temperature ratings impact offset drift, and package constraints affect mechanical stress sensitivity. Understanding how these sections connect is the first step toward reading a datasheet effectively.

Below, we explore the major components of a typical accelerometer datasheet and explain how to interpret them from an engineering perspective. This prepares you for the detailed, parameter-by-parameter discussions in later sections of this chapter.

4.1.1 Device Overview and Key Features

Datasheets typically begin with a brief summary describing the type of accelerometer, which in this book is capacitive MEMS, the available measurement ranges, the output type, whether digital or analog, key performance highlights, and important capabilities such as built-in filters, FIFO support, motion detection, or self-test.

Although this section is short, it provides a quick way to decide whether the device is a potential fit for your project. For example, if you need a $\pm 16\ g$ range with low noise and a low-power wake-on-motion feature, and the overview indicates these capabilities are available, it is worth reading further.

This part of the datasheet functions much like the abstract of a research paper, concise but informative enough to guide an initial decision.

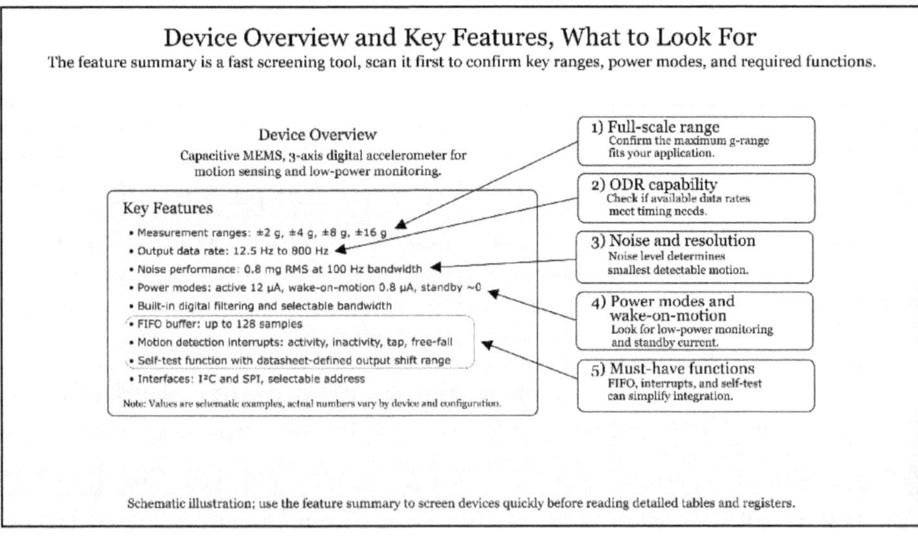

Figure 4.1 Device overview and key features section of an accelerometer datasheet, highlighting what to look for first.

Figure 4.1 illustrates how the device overview and key features section of a datasheet can be used as a rapid screening tool. The highlighted callouts show the specifications engineers typically check first, including full-scale range, ODR capability, noise performance, power modes with wake-on-motion support, and essential features such as FIFO buffering, interrupts, and self-test. Reviewing this summary allows engineers to quickly determine whether a device is a potential fit before investing time in detailed electrical characteristics, timing tables, and register descriptions.

4.1.2 Block Diagram: The Internal Architecture

Most accelerometer datasheets include a block diagram that illustrates the internal architecture, including the MEMS sensing structure, differential capacitors, analog front end, ADC, digital filtering, interrupt logic, I^2C or SPI interface, optional FIFO, and often an internal temperature sensor.

The block diagram is valuable because it shows the signal flow through the device and highlights how measurement data is processed internally. It helps clarify where filtering occurs, whether in the analog or digital domain, whether filtering is applied before or after the ADC, how self-test forces are introduced, whether multiple axes share conversion resources, and how interrupts are generated.

Importantly, understanding the block diagram does not require detailed knowledge of each internal circuit. When reading a datasheet, the primary goal is to identify which functions are implemented inside the sensor, such as digital filtering, FIFO buffering, or temperature sensing, and which tasks must be handled by the host system. This perspective makes it easier to interpret specifications related to bandwidth, ODR, noise behavior, and timing, and helps engineers anticipate configuration options, data latency, and overall integration complexity.

4.1.3 Electrical Characteristics

This section defines electrical limits and operating conditions such as supply voltage (V_{dd}), I/O voltage (V_{ddIO}), current consumption in different operating modes, startup time, input and output logic characteristics, and communication timing for interfaces such as I^2C or SPI.

Supply voltage affects noise performance and stability because inadequate supply level or poor supply quality can degrade analog front-end operation and increase measurement variation, startup time influences power efficiency in duty-cycled systems, communication timing limits achievable data throughput, and I/O voltage compatibility determines whether the device can interface directly with the host microcontroller or requires level translation.

Electrical characteristics are typically presented as tables listing voltage ranges, current consumption, and timing parameters under specific operating conditions. For example, a datasheet may specify a core supply voltage of $1.8\ V\ to\ 3.6\ V$, an I/O voltage range of $1.2\ V\ to\ 1.8\ V$, active-mode current of $120\ \mu A\ at\ 100\ Hz$, and standby current below $1\ \mu A$. When reviewing these tables, it is important to consider the associated operating conditions, such as ODR, bandwidth, and temperature, since current consumption and performance can vary significantly with configuration. Careful interpretation of these specifications helps ensure compatibility with the system power architecture and avoids incorrect assumptions during integration.

4.1.4 Register Map and Configuration Settings

Accelerometers with digital output provide a register map that defines how to enable or disable axes, set the full-scale range, select the ODR, configure bandwidth, enable filters, set interrupt thresholds, select power modes, configure the FIFO, read acceleration data, and run self-test. This register map serves as the software interface to the accelerometer.

Engineers must pay close attention to details such as bit positions, default values, write protection, interrupt routing, multi-byte read requirements, and timing constraints. Misinterpreting even a single register bit can dramatically change sensor behavior; for example, unintentionally enabling a high-pass filter can corrupt tilt measurements.

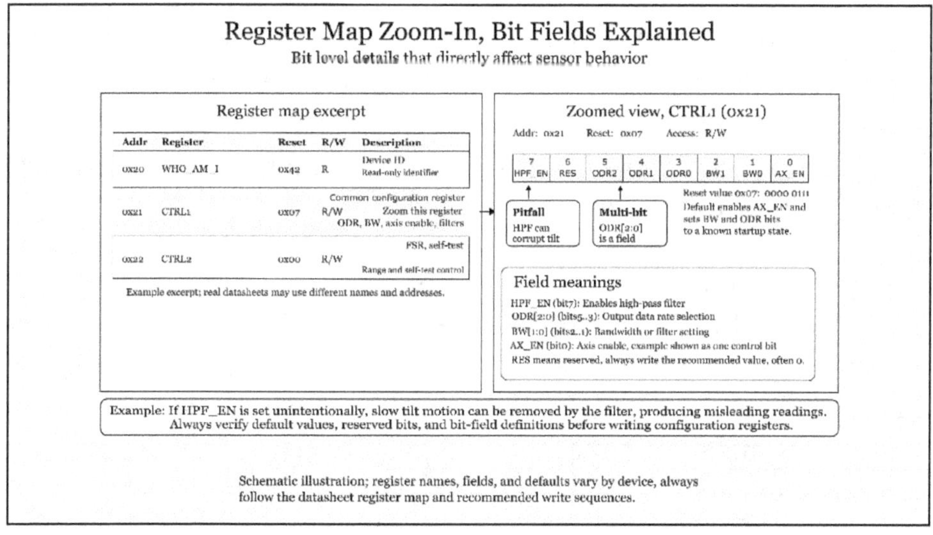

Figure 4.2 Example register map excerpt with a zoomed view of bit fields and their meanings.

Figure 4.2 illustrates how to interpret a register map and its associated bit fields in an accelerometer datasheet. The left panel shows a small excerpt of a typical register table, while the right panel zooms in on a commonly used control register to highlight individual bit positions, default values, access type, and multi-bit fields. The example emphasizes how enabling or disabling a single bit, such as a high-pass filter, can significantly alter sensor behavior. Understanding register structure, default states, and field definitions is essential for configuring the device correctly and avoiding unintended measurement errors.

4.1.5 Mechanical and Package Information

This section includes package drawings, pin assignments, recommended PCB footprints, mechanical tolerances, sensor orientation diagrams, marking information, and land pattern recommendations.

A common engineering mistake is to overlook the orientation diagram, which shows how the X, Y, and Z axes are oriented relative to the package. Mounting an accelerometer rotated by 90° or 180° is acceptable, but requires compensating coordinate transformations in software.

Package information also matters for soldering, reflow temperature profiles, mechanical stress sensitivity, and thermal expansion behavior. Incorrect mounting or inadequate attention to land patterns and orientation markings can introduce offset shifts, axis misalignment, or orthogonality errors that are difficult to diagnose later.

When reviewing this section, engineers should focus on axis orientation, pinout, and recommended land patterns rather than only physical dimensions. Careful attention to these details early in the design process helps prevent integration issues related to mechanical stress, axis alignment, and long-term stability.

4.1.6 Application Notes and Typical Performance Curves

High-quality datasheets are often accompanied by application notes or dedicated sections that explain recommended operating modes, configuration choices, filtering strategies, and example use cases for the sensor. These notes complement the raw specifications by showing how the device is intended to be used in real systems and by highlighting configuration trade-offs that may not be obvious from tables alone.

Datasheets also include performance graphs such as noise versus bandwidth, offset versus temperature, sensitivity drift versus temperature, frequency response, self-test response, and power consumption versus ODR. Unlike numerical tables, these plots reveal trends and interactions between parameters, helping engineers understand how performance changes under different operating conditions.

When reviewing these curves, the goal is not to extract a single best-case value, but to determine whether performance remains acceptable across the intended operating range. For example, a sensor with very low noise at narrow bandwidth may show rapid noise degradation as bandwidth increases, or a device with good room-temperature accuracy may exhibit nonlinear drift at extreme temperature. Interpreting these plots in the context of the application helps engineers identify limitations, anticipate trade-offs, and avoid relying on specifications that only apply under ideal conditions.

4.1.7 Performance Characteristics (The Most Important Section)

This section contains nearly all of the specifications discussed in Chapter 3, including full-scale range, sensitivity, resolution, noise density and RMS noise, bandwidth, ODR, non-linearity, offset, drift, cross-axis sensitivity, shock and vibration limits, self-test response, and temperature behavior.

The performance characteristics section is the core of the datasheet. It provides the key performance parameters that define how the accelerometer is expected to perform under specified conditions and establishes the limits that must be considered during system design. Because these parameters are so important, the remainder of this chapter focuses on how to interpret each one in the context of a real datasheet, not only what each specification means, but how to read it correctly.

This section may be labeled differently depending on the vendor, such as Specifications, Performance Characteristics, or Mechanical Characteristics, but it typically contains the key performance parameters.

4.2 Reading Performance Specifications in a Datasheet

The performance section of an accelerometer datasheet contains the specifications that most engineers focus on when selecting a sensor. These values determine how accurately, consistently, and cleanly the accelerometer measures acceleration. While these specifications were introduced conceptually in Chapter 3, this section explains how to interpret them the way an engineer does when reading an actual datasheet.

Different manufacturers present performance tables in slightly different formats, but the underlying meaning is consistent across devices. A typical table includes columns such as test conditions, minimum, typical, and maximum values, units, notes or footnotes, temperature conditions, and range settings. Understanding these columns is essential, because minimum, typical, and maximum values are not interchangeable, and test conditions often determine whether a specification applies at all.

Below, we walk through each major specification again, this time from the perspective of a real datasheet reader. We explain how manufacturers present the information, how to interpret minimum, typical, and maximum values, what caveats to watch for, and how to extract practical meaning from the numbers.

4.2.1 Full-Scale Range (FSR)

In datasheets, the full-scale range is usually specified in the feature list, the configuration register description, and the performance table under entries such as "Measurement Range." It indicates the maximum acceleration that can be measured along each axis before saturation.

The full-scale range is often selectable through configuration registers. For example, settings may map to ranges such as $\pm 2\ g, \pm 4\ g, \pm 8\ g, or\ \pm$ 16 g, depending on the device. Changing the selected full-scale range also changes the sensitivity, typically by a factor of two for each step.

When evaluating full-scale range options in a datasheet, it is useful to verify how changes in range affect other parameters. Increasing the range usually increases the allowable acceleration before saturation but reduces sensitivity and may influence noise behavior. It is also worth confirming whether sensitivity values scale linearly with the selected range and whether different axes support different full-scale options, which is uncommon but possible in some devices.

4.2.2 Sensitivity

Datasheets define sensitivity using units such as LSB/g or mg per digit for accelerometers with digital output, and mV/g for accelerometers with analog output.

Sensitivity values are typically provided for each selectable full-scale range. For example, a datasheet may list values such as $1024\ LSB/g\ at\ \pm 2\ g, 512\ LSB/g\ at\ \pm 4\ g, 256\ LSB/g\ at\ \pm 8\ g$, and $128\ LSB/g\ at\ \pm 16\ g$. This pattern reflects a fundamental relationship in fixed-resolution accelerometers with digital output: increasing the measurement range reduces the sensitivity.

Because sensitivity depends on the selected full-scale range, datasheets often provide separate tables or entries for each configuration. When reading these tables, it is important to verify which range setting is active, since the same device may produce very different digital output values for the same physical acceleration depending on the selected range.

Sensitivity can also be expressed in mg per digit (mg/LSB), which represents the acceleration corresponding to a single output count. These two representations describe the same physical relationship but from opposite perspectives: LSB/g indicates how many digital counts correspond to one unit of acceleration, while mg/LSB indicates how much acceleration corresponds to one digital count.

Once the digital output code is read from the sensor and converted to a signed integer value, the measured acceleration can be calculated using the sensitivity value provided in the datasheet. The exact register format and numeric representation depend on the device and are described in the datasheet documentation.

4.2.3 Resolution

Datasheets rarely list resolution as a standalone specification. Instead, resolution is inferred from the ADC bit depth, the selected full-scale range, and the resulting sensitivity. As discussed in Chapter 3, resolution is determined by the relationship between full-scale range and the number of available digital output levels.

In many accelerometers with digital output, the datasheet specifies sensitivity values for each selectable full-scale range rather than explicitly stating the ADC resolution. These values implicitly reflect the number of digital codes available across the measurement span.

In practice, this relationship is often interpreted through the sensitivity value given in LSB/g. When sensitivity is expressed in LSB/g, the quantization step size in acceleration units can be obtained by taking the reciprocal of the sensitivity value. For example, a sensitivity of $1024\ LSB/g$ corresponds to a quantization step of $1/1024\ g$ per count.

It is important to recognize that this quantization step represents the ideal digital resolution. In practice, the smallest usable acceleration change is often larger because noise, filtering, and signal-conditioning effects reduce the effective number of usable bits. This behavior is commonly described using the effective number of bits (ENOB).

In practical systems, engineers can improve the effective measurement resolution by selecting an appropriate full-scale range, using sensors with higher ADC resolution, or reducing measurement noise through filtering and averaging.

When evaluating resolution in a datasheet, engineers should consider not only the nominal digital scaling but also the specified noise density and RMS noise, since these parameters ultimately determine the smallest reliably measurable acceleration.

4.2.4 Interpreting Noise and Bandwidth Specifications in Datasheets

Noise density and RMS noise values in datasheets should be interpreted using the foundational definitions and relationships introduced in Chapter 3. Datasheets rarely restate these relationships explicitly; instead, they present noise density as a typical value under defined test conditions and expect the reader to compute total RMS noise based on the selected bandwidth.

When reviewing noise specifications, engineers should pay close attention to the stated bandwidth, power mode, temperature, and full-scale range, as these parameters strongly influence the effective noise seen at the output. Comparing noise values across devices is only meaningful when bandwidth, operating mode, and full-scale range are comparable.

Bandwidth and ODR must also be interpreted together. Datasheets may list multiple ODR settings without explicitly stating the corresponding bandwidth limits. In such cases, the effective bandwidth is determined by the sensor's internal filtering and must remain consistent with the Nyquist condition discussed in Chapter 3 to avoid aliasing.

The key task when reading datasheets is therefore not to re-derive noise or bandwidth theory, but to verify that the published specifications, test conditions, and configuration options align with the system's performance requirements.

4.3 Understanding Accuracy and Error Specifications in a Datasheet

Accuracy-related specifications describe how closely an accelerometer's output matches the true acceleration under various conditions. While Chapter 3 explained these concepts from a physical and engineering perspective, this section focuses on how to read them in a datasheet, interpret minimum, typical, and maximum values, and understand why certain entries matter more than others in real designs.

These parameters are often the most misunderstood part of a datasheet. Engineers who misinterpret offset drift, sensitivity tolerance, or cross-axis effects may end up with systems that behave unpredictably, produce noisy or biased results, or require unnecessary recalibration. Learning how to read these specifications correctly saves time and helps avoid many integration issues.

4.3.1 Zero-g Offset (Offset Error)

In a datasheet, zero-g offset is typically listed in a form such as: zero-g level of $\pm 40\ mg$ typical and $-80\ mg$ to $+80\ mg$ limits (minimum/maximum), measured under defined test conditions such as a full-scale range of $\pm 2\ g$ at $25\ °C$.

The typical value indicates what most devices are expected to exhibit, while the minimum and maximum values represent the worst-case limits guaranteed by the manufacturer. For system design, the minimum and maximum limits must be used rather than the typical value.

If the typical offset is $\pm 40\ mg$ but the minimum/maximum limits are $-80\ mg$ to $+80\ mg$, you should assume the device may output as much as $\pm 0.08\ g$ at true $0\ g$ unless calibration is applied, since the typical value is not guaranteed across all devices. Because offset affects every measurement, it must be characterized during calibration and then compensated in software.

4.3.2 Offset Temperature Drift

Datasheets typically specify offset drift as a zero-g offset temperature coefficient, usually expressed in $mg/°C$ and measured across a specified temperature range. As explained in Chapter 3, this parameter describes how the sensor's zero-g offset changes as temperature varies.

For example, a datasheet may specify an offset drift of $\pm 0.5\ mg/°C$ typical and $\pm 1.0\ mg/°C$ minimum/maximum over a temperature range such as $-40\ °C\ to\ +85\ °C$. These values indicate how much the zero-g output may shift as the device experiences temperature changes.

When interpreting this specification in a datasheet, engineers should estimate the potential offset variation over the expected operating temperature range of the application and verify that the resulting error remains within acceptable limits. Temperature graphs provided in datasheets can also help determine whether the drift is approximately linear or exhibits nonlinear behavior across the operating range.

In many designs, offset drift is reduced through device-level calibration, internal temperature compensation, or system-level correction in software. When evaluating a sensor, it is also useful to check whether the device includes an internal temperature sensor or built-in compensation features that help maintain accuracy across temperature variations.

4.3.3 Sensitivity Tolerance and Sensitivity Drift

Datasheets often specify sensitivity accuracy using a tolerance such as ±3% typical and ±5% minimum/maximum. This means the actual sensitivity may deviate from the nominal sensitivity value listed in the datasheet.

For example, if the nominal sensitivity is specified as 1024 LSB/g, a $\pm$ 5% tolerance indicates that the true sensitivity may vary around that nominal value across devices and operating conditions. When interpreting datasheets, engineers must consider this tolerance when converting digital output codes into physical acceleration values, since sensitivity variation directly affects measurement accuracy.

Sensitivity drift describes how sensitivity changes with temperature. In datasheets, this parameter is usually expressed as a percentage change relative to the nominal sensitivity, either per degree Celsius or across the full operating temperature range. As discussed in Chapter 3, this drift reflects a change in the accelerometer's scaling factor with temperature and can be translated into a corresponding measurement error over the expected operating temperature range.

When evaluating this parameter in a datasheet, engineers should estimate how the specified drift could affect the expected temperature range of the application and determine whether temperature compensation or calibration is required. Devices used in automotive, outdoor, or wearable applications may experience significant temperature variation, making sensitivity drift an important consideration for maintaining measurement accuracy.

4.3.4 Non-linearity

Non-linearity is typically presented in datasheets as a percentage of full scale. This parameter indicates how much the actual output curve may deviate from the ideal straight-line relationship between acceleration and sensor output.

Datasheets often specify non-linearity under defined test conditions and may indicate whether the measurement is performed over the full measurement span or only over one side of the range. In many devices, non-linearity is measured one-sided, from $0\ g$ to the positive full-scale value, rather than across the entire $-FS$ to $+FS$ range. Engineers should verify how the specification is defined, since a one-sided measurement may not fully represent the error behavior in the negative direction.

$$Non-linearity\text{: } 0.2\%\ of\ full\ scale\ (typical)$$

$$Test\ conditions\text{: } measured\ from\ 0\ g\ to\ +FS$$

For example, if the full-scale range is $\pm 4\ g$, then a non-linearity of 0.2% corresponds to:

$$0.2\% \times 4\ g = 0.008\ g = 8\ mg$$

When interpreting non-linearity in a datasheet, it is useful to consider how this deviation may affect system accuracy at higher acceleration levels. Even small percentages of full-scale non-linearity can translate into measurable errors when operating near the limits of the measurement range.

4.3.5 Cross-Axis Sensitivity

Datasheets typically specify cross-axis sensitivity in a form such as 1% typical and 3% maximum. If 1 g is applied along the Y axis, the X axis should ideally report no acceleration. Due to cross-axis sensitivity, however, it may report 0.01 g at the typical level and up to 0.03 g at the maximum level.

Cross-axis sensitivity is not purely an electronic effect; it often arises from mechanical coupling within the MEMS structure. In systems that rely on precise multi-axis measurements, such as drone stabilization or tilt sensing, cross-axis sensitivity can introduce noticeable errors if not properly accounted for.

4.3.6 Alignment Error (Axis Misalignment)

Datasheets sometimes specify axis misalignment, for example $\pm 2°$ typical and $\pm 3°$ maximum. This parameter describes how much the physical sensing axes deviate from perfect orthogonality.

Axis misalignment is important in applications such as robotics, drones, navigation, and multi-axis tilt calculations. If the misalignment angle is known, it can often be corrected in software.

4.3.7 Noise Density and RMS Noise

Section 4.2 explained how noise specifications are presented in datasheets. It is also important to understand how to interpret noise density and RMS noise values in practice, including the effects of bandwidth, power mode, and filtering.

In datasheets, noise density is usually listed as a typical value under defined test conditions and may not be guaranteed across the full operating range. When reading these entries, always note the associated bandwidth, power mode, supply voltage, and temperature conditions listed in the table footnotes.

Some datasheets also list RMS noise directly for specific bandwidth settings, while others require the user to compute RMS noise from the noise density and the selected bandwidth. When comparing devices, ensure that noise values are evaluated under comparable conditions and configurations, as changes in bandwidth, power mode, or filtering can significantly affect noise performance.

4.3.8 Hysteresis

Datasheets may list hysteresis values such as ±15 mg typical. This specification represents the difference in sensor output for the same acceleration depending on whether the input is increasing or decreasing, as explained in Chapter 3.

In most modern MEMS accelerometers, hysteresis is very small and often negligible for dynamic measurements. However, it can become noticeable in applications that rely on precise static or slowly changing accelerations, such as tilt sensing or other DC measurements.

When evaluating a datasheet, hysteresis should be considered together with offset and long-term stability, since these parameters collectively determine the repeatability of measurements when acceleration conditions change over time.

4.3.9 Total Error Band (TEB) or Overall Accuracy

Some manufacturers provide a TEB specification defined as the combined contribution of multiple error sources across temperature and operating range. This value may include offset error, offset drift, non-linearity, sensitivity error, sensitivity drift, noise contributions, and hysteresis. As explained in Chapter 3, these individual error sources can be combined to form a conservative worst-case accuracy bound for the accelerometer.

TEB gives a convenient worst-case estimate under the manufacturer's stated conditions. Because this specification aggregates several individual error sources, it offers a convenient way to estimate worst-case system accuracy without analyzing each parameter separately.

When interpreting TEB in a datasheet, engineers should carefully examine the test conditions and assumptions used to derive the value. TEB specifications often assume particular bandwidth or filtering settings, and the effective error may change if the device is operated under different configurations.

For this reason, the stated TEB should be treated as a worst-case estimate rather than an exact prediction of system performance. Engineers should verify that the conditions under which the specification is defined match the intended operating environment of the application.

4.4 Interpreting Electrical and Operational Specifications

Electrical and operational specifications describe how the accelerometer interacts with the rest of the system. These parameters determine how the device is powered, how it communicates, how much current it draws, how long it takes to start up, and how its built-in features behave. While performance specifications define measurement quality, electrical specifications determine whether the device functions correctly in a given design and how efficiently it operates.

These specifications are often underappreciated by beginners, but in practice they have a major impact on system behavior, especially in battery-powered, timing-critical, or safety-related designs. Understanding how to interpret the electrical section of the datasheet helps ensure that the accelerometer integrates smoothly with the microcontroller, power supply, PCB, and software.

4.4.1 Supply Voltage (V_{dd}) and I/O Voltage (V_{ddIO})

As discussed in Chapter 3, many accelerometers use separate supply domains for the sensor core and the digital interface to support compatibility with different logic levels. Most datasheets specify two voltage domains: V_{dd}, which powers the MEMS structure and ASIC, and V_{ddIO}, which defines the logic-level voltage used for communication interfaces such as I^2C or SPI. An example entry may list V_{dd} from 1.6 V to 3.6 V and V_{ddIO} from 1.2 V to 3.6 V.

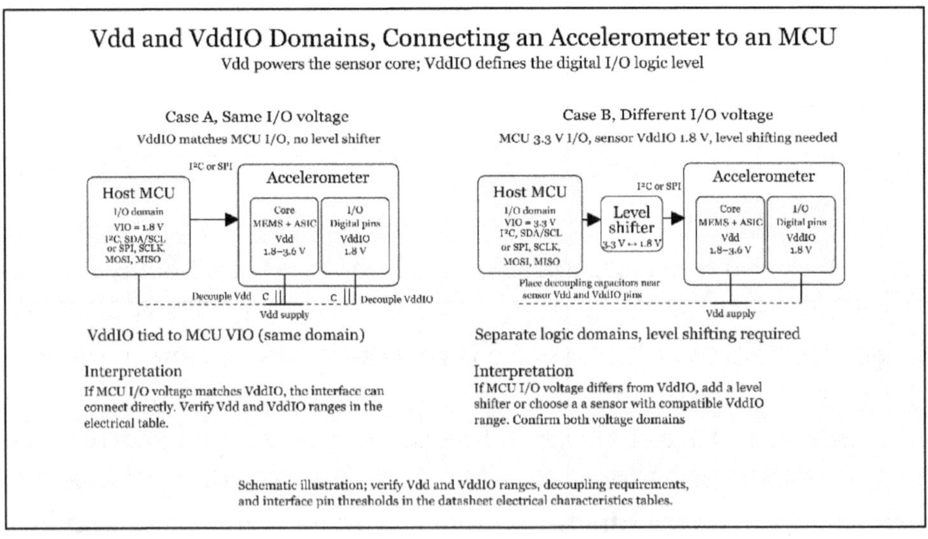

Figure 4.3 V_dd and V_ddIO voltage domains when connecting an accelerometer to a microcontroller.

Figure 4.3 illustrates how digital accelerometers often use two separate supply domains: V_{dd}, which powers the sensor core including the MEMS structure and internal circuitry, and V_{ddIO}, which defines the logic-level voltage for the I^2C or SPI interface. When the microcontroller I/O voltage matches V_{ddIO}, the interface can be connected directly. If the MCU uses a different logic level, a level shifter is required to ensure reliable communication. The capacitors shown near the V_{dd} and V_{ddIO} pins represent local decoupling, which stabilizes each supply domain, reduces noise coupling, and ensures proper operation of both the sensor core and the digital interface. Datasheet electrical characteristics should always be consulted to confirm allowed voltage ranges and decoupling requirements.

When reviewing these specifications, engineers should verify logic-level compatibility between the sensor and the MCU, confirm that adequate decoupling is used near the sensor, and avoid assuming that noise performance is identical across all V_{dd} values. Some sensors also allow V_{ddIO} to be lower than V_{dd}, which can reduce I/O switching power.

4.4.2 Current Consumption (Operating Modes)

Datasheets list current consumption for multiple operating modes, such as active mode, low-power mode, wake-on-motion, and standby. Typical values might include 10 μA in active mode, 2 μA in low-power mode, 0.8 μA in wake-on-motion, and 0.2 μA in standby.

Each operating mode trades performance for power savings. Active mode provides full bandwidth and ODR, while low-power modes typically save power by reducing one or more performance-related functions, such as bandwidth, ODR, or internal signal processing activity. Wake-on-motion modes keep only minimal circuitry active so the device can detect movement and trigger an interrupt.

When reviewing current consumption tables, always cross-check the associated bandwidth, ODR, and noise specifications for each mode. Datasheets may list attractive low-power currents without clearly emphasizing the corresponding reduction in bandwidth or increase in noise, which can significantly affect usable performance.

4.4.3 Startup Time (Turn-On Time or Power-Up Time)

Datasheets specify startup time as the interval required for the accelerometer to become operational after power is applied. For example, a device might list a startup time of 5 *ms* typical and 10 *ms* maximum. This specification indicates how long the internal circuitry needs to stabilize before valid measurements can be produced.

When interpreting this parameter in a datasheet, engineers should verify whether the specified time refers only to the device power-up or also includes the time required for filters, ODR settings, or internal signal processing to settle. Some devices may require additional time before measurements become fully stable.

Startup time becomes particularly important in duty-cycled systems where the sensor is frequently powered down to save energy. In such cases, the startup delay directly affects system responsiveness and overall power efficiency.

4.4.4 Communication Interface Specifications (I^2C / SPI)

Datasheets include timing diagrams and electrical characteristics that define how the accelerometer communicates with a host processor. These specifications typically include setup and hold times, the maximum I^2C Serial Clock Line (SCL) frequency, SPI clock polarity and phase modes, bus voltage thresholds, multi-byte read behavior, and device addressing schemes.

The maximum I^2C clock frequency defines the fastest achievable data readout. The selected SPI clock polarity and phase setting, commonly identified as mode 0, 1, 2, or 3, must match the MCU configuration. I^2C interfaces use open-drain signaling and therefore require appropriate pull-up resistors. When reviewing timing diagrams, focus on maximum clock rates, setup and hold times, and multi-byte read behavior rather than waveform shape, since an accelerometer configured with an incorrect SPI mode or missing I^2C pull-up resistors will not communicate reliably.

4.4.5 Self-Test Behavior and Specifications

Datasheets specify the expected output change when the accelerometer's built-in self-test function is enabled. For example, a datasheet may specify a self-test response on the X axis in the range of $+0.25\ g$ to $+0.45\ g$.

During self-test, the measured output should fall within the specified range. If the response lies outside these limits, it may indicate a fault in the device, such as damage to the MEMS structure or malfunction in the associated analog or digital circuitry. In some applications, particularly automotive or medical systems, periodic self-test is used to verify sensor integrity during operation.

The self-test mechanism itself is illustrated in **Figure 3.12**, which shows how an internal electrostatic force deflects the proof mass and produces a predictable output shift. As discussed in Chapter 3, the key task when interpreting a datasheet is simply to verify that the measured self-test response falls within the guaranteed minimum and maximum limits under the specified test conditions.

4.4.6 Interrupt Behavior and Threshold Registers

Many datasheets include details about motion interrupts, tap or double-tap detection, free-fall interrupts, wake-on-motion thresholds, slope detection, and the filtering paths used for interrupt generation.

Interrupt logic often uses filtering paths that are separate from the main acceleration data path. For example, a motion interrupt may use a 25 Hz filter while the acceleration data is produced using a 100 $Hz\ filter$. This behavior is common and explains why interrupts may appear to trigger earlier or later than expected when compared with raw data. As a result, interrupt thresholds should not be assumed to correspond directly to raw acceleration values, since interrupt latency depends on the selected ODR and bandwidth, and overly low thresholds can allow noise in the signal to produce false triggers, especially in noisy environments.

4.4.7 FIFO and Data Buffering Specifications

FIFO sections describe how sample data is stored in the buffer, including whether each entry contains a complete X-Y-Z sample or whether X, Y, and Z values are stored as separate sequential entries. They also list parameters such as maximum depth, watermark interrupt levels, available FIFO modes such as bypass, stream, or trigger, and timing guarantees. FIFO buffering reduces MCU interrupt load, ensures consistent timing, enables burst reads, and improves power efficiency.

Error! Reference source not found. illustrates how an internal FIFO (First-In, First-Out) buffer allows an accelerometer to continue sampling while the microcontroller is busy or asleep. In the timeline view, samples are produced continuously at the selected ODR and accumulated in the FIFO until a programmable watermark level is reached, at which point an interrupt can wake the MCU to perform a burst read. The FIFO stack view shows how samples are stored internally, typically as XYZ triplets ordered in time, and how the watermark threshold defines when data should be retrieved. This mechanism reduces MCU wake-ups, improves power efficiency, and prevents data loss in systems where the MCU cannot read every sample immediately.

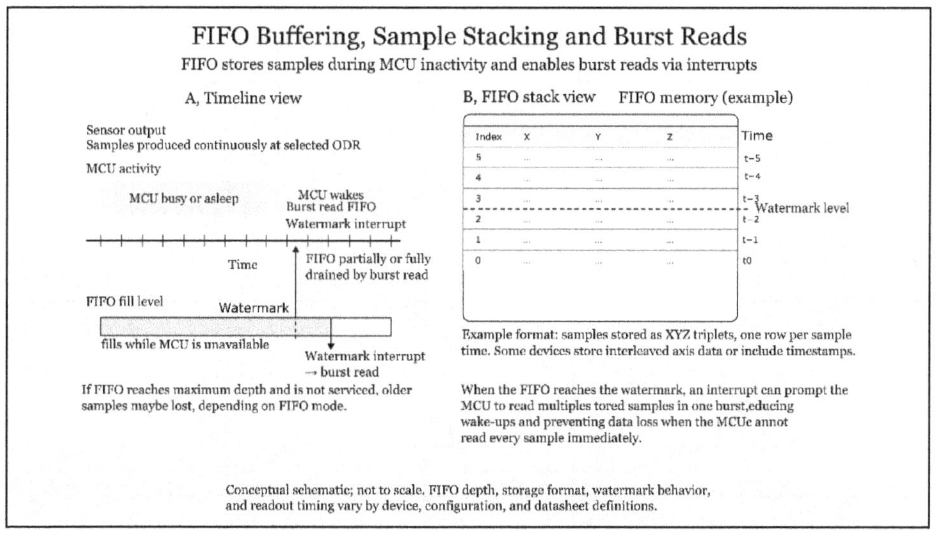

Figure 4.4 FIFO buffering, sample stacking, and burst reads in a digital accelerometer.

In many systems, the microcontroller is not continuously available to read sensor data as soon as it is produced. The MCU may be busy handling other tasks, servicing peripherals, or operating in a low-power state. In other cases, designers intentionally choose to buffer sensor data and read it in bursts to reduce wake-ups and save power. A FIFO prevents data loss in both situations by allowing the accelerometer to store samples internally until the system is ready to retrieve them.

4.4.8 Filtering Options (Analog and Digital)

Datasheets specify filtering features such as analog low-pass filter cutoff, available digital filter options, high-pass filter behavior, and selectable bandwidth modes.

It is important to note whether analog and digital filters are cascaded (applied in sequence), which filter path is used for interrupt detection, and whether a high-pass filter removes the DC gravity component, which is critical in tilt applications. When multiple filter paths are present, datasheets typically specify which path applies to the output data and which applies to interrupt generation; these paths are often different. Leaving a high-pass filter enabled can make tilt measurements appear noisy or unstable, since the filter removes the DC gravity component.

4.5 Understanding Environmental and Reliability Specifications

Environmental and reliability specifications describe the conditions under which an accelerometer can operate safely and reliably. These parameters are especially important for automotive, industrial, medical, wearable, and outdoor applications, where temperature variation, shock events, vibration, humidity, and mechanical stress can all influence sensor behavior.

Datasheets typically list environmental specifications near the end of the performance tables or in a dedicated environmental section. These values define absolute limits rather than recommended operating points, and exceeding them can lead to permanent damage or unpredictable behavior.

4.5.1 Operating Temperature Range

Datasheets specify the operating temperature range over which the accelerometer is guaranteed to meet its performance specifications. Typical ranges include $-40\ °C$ to $+85\ °C$ for many consumer and general-purpose industrial devices, while extended ranges such as $-40\ °C$ to $+105\ °C$ or $-40\ °C$ to $+125\ °C$ are common in harsher industrial and automotive applications.

The device is guaranteed to meet all electrical and performance specifications only within the stated operating range. Performance outside this range is not guaranteed, and drift effects generally increase near the temperature extremes.

When interpreting this specification in a datasheet, engineers should verify whether temperature-dependent graphs are provided for parameters such as offset drift, sensitivity drift, or noise. It is also important to distinguish between operating temperature range and storage temperature limits, which may be different. As discussed in Chapter 3, temperature variations influence several sensor parameters, so the specified operating range should be compared carefully with the expected environmental conditions of the application.

4.5.2 Shock Resistance

Datasheets typically specify shock resistance as the maximum acceleration the device can survive without permanent damage. This value is usually given as a peak acceleration applied over a short duration, for example 10,000 g for 0.1 ms. The specification indicates that the sensor can withstand extreme mechanical shocks such as drops, collisions, or vibration bursts.

When interpreting this parameter, it is important to distinguish between shock survival limits and the measurement range of the accelerometer. A device rated to survive 10,000 g shock may still only measure accelerations accurately within a much smaller range, such as $\pm 2\ g$ or $\pm 16\ g$.

Shock resistance is particularly relevant for devices exposed to mechanical impacts, including smartphones that may be dropped, industrial sensors installed in harsh environments, automotive electronics subjected to crash pulses, and drones experiencing hard landings. During such events the internal proof mass may contact mechanical stops, but the sensor remains structurally intact as long as the applied shock stays within the specified limits.

4.5.3 Vibration Resistance

Datasheets often specify vibration limits using entries such as 20 g RMS from 20 Hz to 2000 Hz, or sine vibration of $\pm 2\ g$ from 10 Hz to 1000 Hz. These specifications indicate the vibration levels the sensor can tolerate during continuous excitation, the frequency range over which the device was tested, and whether the test used random vibration (expressed as RMS values) or sine-wave vibration.

When reviewing vibration specifications, engineers should verify the test conditions and frequency range used to define the limit. Vibration ratings are typically intended to ensure mechanical survivability and stable operation rather than to describe the detailed measurement response of the sensor under vibration.

Datasheets may also provide frequency response plots showing the sensor's behavior across frequency. These plots should be examined to ensure that the intended operating frequencies remain well below the sensor's resonant frequency, where vibration can distort measurements.

Applications that commonly require strong vibration tolerance include industrial machinery, vehicle powertrains, drones, and heavy equipment operating in dynamic environments.

4.5.4 Humidity, Moisture, and Environmental Sealing

Datasheets may specify humidity limits such as $0\% \ to \ 95\%$ relative humidity (non-condensing) and may also note the presence of ESD-protected pads, moisture sealing, or optional hermetic packaging.

When reviewing humidity specifications, engineers should note that values such as 85% to 95% relative humidity are typically defined as non-condensing and measured under moderate temperature conditions (for example, around $40\ °C$ to $60\ °C$ ambient temperature). These limits do not imply suitability for condensation, immersion, or prolonged exposure to liquid moisture.

Applications exposed to outdoor environments, rapid temperature cycling, or sustained high humidity should therefore consider whether additional environmental protection is required at the system level, such as sealed housings, protective coatings, or conformal coating on the circuit board.

4.5.5 Mechanical Stress Sensitivity

Datasheets may include specifications describing how sensitive the accelerometer is to mechanical stress applied through the package or printed circuit board (PCB). These parameters may appear as values such as offset shift versus board bending (for example, 0.01 g per millimeter of PCB deflection) or package stress sensitivity expressed in mg (milli-g).

Such specifications indicate how much the sensor output may change when the PCB flexes, mounting screws are tightened, or mechanical pressure is applied to the package. Because the sensor die is mechanically coupled to the package and PCB, external stress can alter the alignment of the MEMS structure or change internal capacitances, resulting in measurable offset shifts.

When interpreting these specifications in a datasheet, engineers should consider the expected mechanical stresses in the final system and verify that the resulting output shifts remain acceptable. Sensors used in precision applications should be mounted on rigid regions of the PCB, away from screw holes or flex points, and the recommended land pattern and soldering guidelines should be followed to minimize stress-induced errors.

4.5.6 ESD Ratings (HBM, CDM)

Datasheets specify electrostatic discharge protection using industry-standard models such as the Human Body Model (HBM) and the Charged Device Model (CDM). Typical ratings may range from about $\pm 2\ kV$ to $\pm 4\ kV$ for HBM and from about $\pm 500\ V$ to $\pm 1000\ V$ for CDM, though values vary by device and package.

Figure 4.5 compares the two most common electrostatic discharge models used in accelerometer datasheets. HBM represents a discharge from a person to the device during handling or assembly, while CDM represents a device that becomes charged and then discharges rapidly to ground through its package pins.

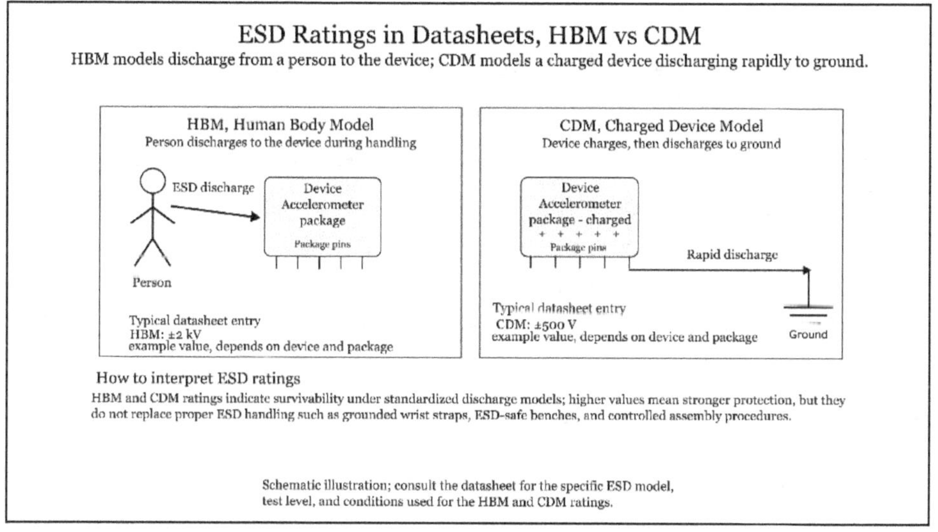

Figure 4.5 Electrostatic discharge models used in accelerometer datasheets: HBM versus CDM.

HBM simulates discharge from human contact during handling or assembly, while CDM represents charge buildup on the device itself followed by a rapid discharge. Higher ESD ratings indicate stronger protection, but they do not eliminate the need for proper handling procedures.

Electrostatic discharge can degrade sensor performance without causing immediate failure, leading to increased noise, offset shifts, or reduced long-term reliability. During assembly and testing, engineers should use grounded wrist straps, ESD-safe work surfaces, and appropriate handling practices to protect sensitive MEMS accelerometers.

Datasheets typically list ESD ratings for both models under standardized test conditions. Higher ratings indicate stronger on-chip protection, but they do not eliminate the need for proper ESD handling practices during manufacturing, testing, and system integration.

4.5.7 Reliability Validation and Endurance Tests

High-quality datasheets often include information about reliability validation and endurance testing, such as temperature cycling, high-temperature operating life testing, mechanical shock testing, vibration endurance, humidity exposure, and drop testing. These tests indicate that the device has been subjected to long-duration environmental and mechanical stress intended to reveal wear-out mechanisms, material fatigue, and long-term drift before the product is released.

When reviewing this section, check whether the device conforms to recognized standards such as AEC-Q100, ISO, or JEDEC, and whether it is qualified as automotive-grade, for example Grade 2 or Grade 1. Datasheets may also include notes on long-term offset drift, sensitivity stability, or material fatigue observed during testing. Sensors that meet automotive or industrial qualification standards have undergone significantly more rigorous validation than typical consumer-grade devices, making them better suited for safety-critical or long-lifetime applications.

4.6 From Raw Output to Final Acceleration Estimate

Converting accelerometer output into a final acceleration value is a multi-step process. The starting point is the raw digital output for each axis, read from the device registers and converted into a signed integer according to the data format defined in the datasheet. This raw count is then converted into an initial acceleration estimate using the nominal sensitivity provided in the datasheet. If sensitivity is expressed in g/LSB or mg/LSB, the raw count is multiplied by that value. If sensitivity is expressed in LSB/g, the raw count is divided by the sensitivity value. If the device output has already been filtered internally, that estimate reflects the filtered signal rather than the raw sensor response.

The next step is to apply any known calibration corrections. Zero-g offset is often the first correction, because even when no acceleration is applied along an axis other than gravity, the output may not be exactly zero. If calibrated sensitivity or scale-factor correction is available, that should also be applied instead of relying only on the nominal datasheet value. In applications that operate across a wide temperature range, offset drift and sensitivity drift may also need to be compensated using temperature data from the sensor or from system-level calibration. These steps improve the converted value so that it more closely represents the true applied acceleration under actual operating conditions.

After these direct corrections are applied, other specifications must still be considered because they affect the remaining accuracy of the result. Non-linearity describes how far the sensor response may deviate from an ideal straight-line transfer curve, especially toward the limits of the selected full-scale range. Cross-axis sensitivity and axis misalignment can introduce additional error when acceleration is present on multiple axes or when the sensor orientation is not perfectly aligned with the system coordinate frame. Hysteresis may cause slightly different outputs for the same acceleration depending on whether the input is increasing or decreasing. These effects are usually not corrected by a simple single-value adjustment, but instead are treated as residual error sources that limit final accuracy.

Noise must also be treated differently from fixed error terms. Noise density and RMS noise do not create a constant offset that can simply be added or subtracted; instead, they determine the uncertainty and repeatability of the measurement. For that reason, the final result should be viewed as a best acceleration estimate together with an expected uncertainty range. In practice, engineers may use the datasheet's total error band when it is available, or they may combine the relevant individual terms, such as offset, drift, non-linearity, hysteresis, and noise, to establish a worst-case or application-specific error bound. The final acceleration value is therefore not just a converted register reading, but a processed and interpreted result whose accuracy depends on calibration, operating conditions, and the relevant datasheet specifications.

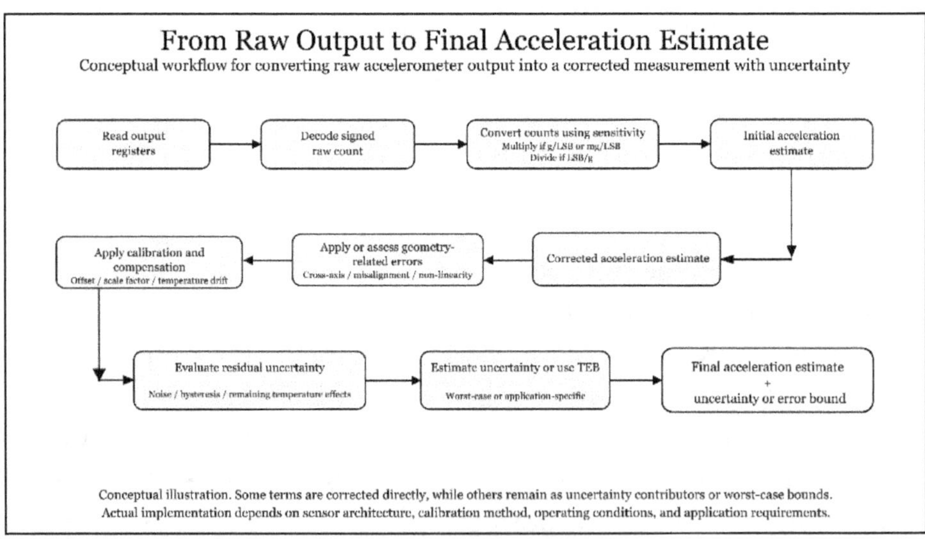

Figure 4.6 Conceptual workflow from raw accelerometer output to a final acceleration estimate with uncertainty.

Figure 4.6 summarizes the practical steps used to convert raw accelerometer output into a final interpreted measurement. The workflow begins with register reading, signed-count decoding, and sensitivity-based conversion, then applies calibration and compensation terms such as offset, scale factor, and temperature drift. Geometry-related effects, including cross-axis sensitivity, misalignment, and non-linearity, are then considered before evaluating residual uncertainty from noise, hysteresis, and remaining temperature effects. The result is not just a corrected acceleration estimate, but a final value together with an uncertainty or error bound.

4.7 Takeaways

In this chapter, we developed one of the most practical skills an engineer can have when working with accelerometers: the ability to read and interpret a datasheet with confidence. A datasheet is more than a list of specifications; it is a compact description of a sensor's physical behavior, electrical constraints, performance characteristics, error sources, and environmental limits. Understanding how to interpret these values is essential for selecting the right accelerometer, predicting system performance, and avoiding design mistakes.

We began by examining the overall structure of a typical datasheet. The feature summary, block diagram, electrical tables, performance specifications, register maps, mechanical drawings, application notes, and package information each serve a distinct purpose. Together, they describe how the accelerometer works internally, how it should be configured, and how it will behave in a real application. Recognizing this structure makes the rest of the datasheet much easier to interpret.

Next, we learned how to read performance tables, focusing on measurement-related specifications such as full-scale range, sensitivity, resolution, noise density, RMS noise, bandwidth, and

ODR. These parameters determine output quality. We then examined accuracy and error specifications, including offset, drift, non-linearity, cross-axis sensitivity, and alignment error. Understanding how manufacturers define minimum, typical, and maximum values, and how these depend on test conditions, allows engineers to assess true performance under realistic operating conditions.

We also explored electrical and operational specifications that govern how the accelerometer interacts with the rest of the system, including power consumption, supply voltage, startup time, digital interface requirements, self-test behavior, interrupt logic, filtering paths, and FIFO buffering. These details are just as important as measurement accuracy because they determine whether the sensor can be integrated reliably into the final system. Finally, we reviewed environmental and reliability specifications that describe how well the sensor withstands temperature extremes, vibration, humidity, shock, and electrostatic discharge, which are critical considerations for automotive, industrial, and outdoor systems.

We also examined how raw output data is transformed into a final acceleration estimate. This process begins by reading the sensor registers and converting the raw digital count using the datasheet sensitivity, then applying available calibration and compensation terms such as offset, scale factor, and temperature correction. Remaining effects, including non-linearity, cross-axis sensitivity, misalignment, hysteresis, and noise, must then be treated as residual error sources or uncertainty contributors. This final step is important because it connects individual datasheet specifications to the actual acceleration value used in a real system.

To tie everything together, the checklist below summarizes a practical method for evaluating an accelerometer datasheet. This compact workflow can be used when comparing devices, reviewing a new sensor, or preparing for design reviews and technical interviews.

With a solid understanding of how to read a datasheet, we now turn to a topic that is just as important: common mistakes engineers make when selecting, integrating, and interpreting accelerometers. Chapter 5 focuses on these pitfalls and shows how to design systems that fully leverage the capabilities of modern MEMS accelerometers.

CHAPTER 5 Common Mistakes with Accelerometers

Even experienced engineers encounter problems when working with accelerometers. These sensors are mechanically and electrically complex, and their real-world behavior depends not only on the MEMS structure and ASIC, but also on how they are mounted, powered, filtered, sampled, and interpreted. Many issues arise not because the accelerometer is poorly designed, but because the data is misread, the configuration is incorrect, or the system environment introduces unintended effects. Understanding the most common mistakes helps prevent wasted development time, confusing results, and unreliable system performance.

In previous chapters, we examined what accelerometers measure, how they work, how specifications are defined, and how to read a datasheet correctly. This chapter builds on that foundation by focusing on the practical mistakes engineers encounter when selecting, integrating, and using accelerometers in real systems. Many of these mistakes stem from misunderstandings about noise, bandwidth, range selection, power modes, axis orientation, filtering choices, and the effects of temperature and mechanical stress.

This chapter is intentionally practical and focused on real-world behavior. Its goal is not to introduce new theory, but to help you recognize common failure modes before they appear in the field. By understanding these mistakes, you will be better equipped to design reliable, accurate, and efficient systems that use accelerometer data.

5.1 Choosing the Wrong Full-Scale Range

One of the most common mistakes engineers make when working with accelerometers is selecting an inappropriate full-scale range, or FSR. Although this decision may seem minor, it directly affects resolution, noise performance, saturation behavior, and overall system accuracy. Incorrect range selection often leads to data that is either too coarse to be useful or that clips during normal operation.

The full-scale range defines the maximum measurable acceleration along each axis. For example, a $\pm 2\ g$ accelerometer measures acceleration between $-2\ g$ and $+2\ g$. If the true acceleration exceeds this range, the output saturates and the measurement becomes invalid. Conversely, selecting a range that is much larger than required reduces sensitivity, making small motions difficult to distinguish from noise.

Selecting the correct range requires understanding the application's expected motion, including peak accelerations, transient events, and safety margins, as well as the trade-off between resolution and saturation.

Selecting a full-scale range that is much larger than necessary is a common mistake that quietly degrades accelerometer performance. While a wide range may seem safer, it spreads the sensor's finite digital resolution over a larger acceleration span, reducing sensitivity and making small motions harder to distinguish from noise.

Accelerometers with digital output allocate a fixed number of digital output levels (counts) across the selected full-scale range. Increasing the range therefore reduces the number of counts per

g, even though the physical motion being measured has not changed.

Although the datasheet already provides the sensitivity for each selectable full-scale range, it is useful to see how those values arise because they show how the same number of digital output levels is distributed across different measurement spans. This also explains why selecting a larger full-scale range increases the acceleration represented by each digital step and makes small motions harder to resolve.

Consider a 14-bit accelerometer with 16,384 ($= 2^{14}$) digital output levels (counts):

With $\pm 2\ g$ selected, the total span is $4\ g$

$$\rightarrow \mathit{Ideal\ resolution} = 4 \,/\, 16{,}384 \approx 0.000244\ g\ \mathit{per\ count} \\ = 0.244\ mg/LSB$$

With $\pm 16\ g$ selected, the total span is $32\ g$

$$\rightarrow \mathit{Ideal\ resolution} = 32 \,/\, 16{,}384 \approx 0.001953\ g\ \mathit{per\ count} \\ = 1.953\ mg/LSB$$

The $\pm 2\ g$ setting provides a digital step size that is eight times smaller than the $\pm 16\ g$ setting. This means small acceleration changes are represented more finely, making them easier to distinguish from quantization limits and noise. If the application never experiences accelerations above $\pm 2\ g$, using $\pm 16\ g$ spreads the same number of digital output levels over a much larger span, reducing useful detail for small motions.

This loss of effective resolution matters in applications that rely on detecting subtle changes, such as tilt sensing, posture monitoring, gait analysis, step counting, and low-frequency motion tracking. In these cases, unnecessarily large ranges lead to jittery outputs and reduced measurement precision, even though the sensor itself is functioning correctly.

Selecting a full-scale range that is too small causes a different but equally serious problem: signal saturation. When the applied acceleration exceeds the selected range, the accelerometer output clips at its maximum or minimum value, permanently losing information about the true motion.

Once saturation occurs, no amount of filtering or post-processing can recover the missing data. The accelerometer is no longer measuring acceleration; it is reporting its own limits.

Assume an accelerometer is configured for $\pm 2\ g$, but the true acceleration reaches higher values during operation. If the actual motion sequence is:

$$0.0\ g, 0.4\ g, 1.1\ g, 1.8\ g, 2.4\ g, 3.0\ g$$

the reported output becomes:

$$0.0\ g, 0.4\ g, 1.1\ g, 1.8\ g, 2.0\ g, 2.0\ g$$

All values beyond $\pm 2\ g$ are clipped. The true peak acceleration and its shape are lost, and the resulting waveform no longer represents the real motion.

This problem is especially damaging in applications with rapid or impulsive motion, such as drones, robotics, vehicles, sports motion tracking, and industrial machinery. Saturation flattens peaks, distorts timing, and can destabilize control loops or corrupt motion classification algorithms.

Even brief saturation events can cause lasting issues. A single clipped peak may bias statistical measures, break frequency analysis, or trigger incorrect decisions in threshold-based systems.

To avoid saturation, the selected full-scale range must exceed the maximum expected acceleration, including transient events and safety margin. Choosing a range that is just barely sufficient leaves no tolerance for unexpected motion and increases the risk of clipped data.

A reliable engineering rule is to use the smallest full-scale range that will not saturate under worst-case conditions. In practice, this means estimating the maximum expected acceleration, applying a safety factor, and selecting the next available range that comfortably exceeds the result.

If a handheld device experiences peak accelerations of approximately 1.8 g during rapid movement, applying a safety factor of 2 yields a required range of about 3.6 g. The appropriate selection is therefore $FSR = \pm 4\ g$.

Incorrect range selection often shows up as poor system behavior rather than an obvious configuration error. In wearable step counters, an excessively large range causes small, repetitive motions to sink into noise, leading to missed steps or erratic counts. The sensor is working, but the resolution is too coarse to capture subtle movement reliably.

In industrial vibration monitoring, the opposite mistake is common. A range that is too small saturates during high-amplitude events, flattening vibration peaks and corrupting frequency-domain analysis. Once clipping occurs, important information about vibration intensity and frequency content is lost, making condition monitoring unreliable.

In consumer tilt sensing and orientation tracking, overly large ranges reduce sensitivity and increase jitter in gravity measurements. This produces unstable angle estimates and a device that feels imprecise or slow to settle, even when the hardware quality is high.

These issues are often misdiagnosed as noise problems, algorithm flaws, or sensor defects. In many cases, the root cause is simply a poorly chosen full-scale range that does not match the real motion of the application.

Start by understanding the full motion envelope of your application before selecting the full-scale range. This includes not only normal operation, but also brief peaks caused by impacts, rapid movement, or user interaction.

Test the expected peak accelerations early in development rather than relying only on assumptions or lab simulations. Early testing often reveals acceleration peaks that are easy to miss during desk analysis.

After selecting an initial range, revisit the decision once noise and resolution requirements are better understood. If small motions appear noisy or unstable, the range may be larger than necessary. If clipping appears during testing, the range is too small.

When applications operate in multiple modes, such as gentle motion most of the time with occasional high acceleration events, sensors that support dynamic range switching can be useful. In these cases, remember that changing the range also changes sensitivity, so software scaling and thresholds must be updated accordingly.

5.2 Misunderstanding Noise

This section assumes familiarity with the noise and bandwidth concepts defined in Chapter 3 and focuses on how misconfiguration or misinterpretation of those concepts leads to real-world failures.

Noise is one of the most misunderstood aspects of accelerometer performance. Many engineers interpret noise as faulty data or assume it can be eliminated entirely through filtering. Others misread datasheet specifications and expect perfectly stable outputs when the device is at rest. These misunderstandings often lead to wasted debugging time, incorrect assumptions about sensor failure, and poor design decisions.

Noise is not a defect. It is a fundamental physical characteristic arising from the MEMS structure, the analog front end, and the digital conversion process. Thermal motion at microscopic scales, electronic charge fluctuations, quantization effects, and internal clock behavior all contribute. The goal is not to eliminate noise, but to understand it, predict it, and design systems that tolerate it.

A very common mistake is assuming that an accelerometer is defective because its output fluctuates while the device is motionless. Small variations in the readings are normal and expected, even for high quality sensors.

An accelerometer resting on a table may report

$$a_x = +0.006\,g, \qquad a_y = -0.004\,g, \qquad a_z = +1.012\,g$$

These small changes do not indicate a problem. They are the natural result of sensor noise and digital representation. Several effects contribute to this behavior. Random output variation is often described by the sensor's RMS noise over the selected bandwidth. ADC quantization introduces discrete step size limits in the digital output, while temperature-related drift and electrical disturbances can add further variation. Axes measuring near 0 g often appear noisier because the true signal is small compared with the combined uncertainty and noise level. Expecting a perfectly constant output is unrealistic. A stationary accelerometer will always show some jitter, and this does not mean the sensor is damaged or poorly designed.

Noise behavior changes with temperature as well. Noise density typically increases as temperature rises, although this effect is often understated in datasheets. Noise density may increase from $150 \frac{\mu g}{\sqrt{Hz}}$ at 25 °C to $200 \frac{\mu g}{\sqrt{Hz}}$ at 85 °C. At a bandwidth of 50 Hz, RMS noise increases noticeably, pushing algorithms closer to their noise limits. Thresholds tuned at room temperature may fail at elevated temperatures, producing missed detections or false triggers.

Figure 5.1 shows how accelerometer noise density typically increases with temperature. As temperature rises, changes in silicon properties, MEMS damping, and analog circuit behavior raise the noise floor. For a fixed bandwidth, this increase directly translates into higher RMS noise, reducing signal-to-noise ratio and shrinking detection margins. Thresholds tuned at room temperature may therefore fail at elevated temperatures, leading to missed events or false triggers unless temperature effects are considered.

A common source of confusion is treating noise density as if it represents the total noise of the accelerometer. In reality, noise density only describes how much noise exists in a very narrow frequency slice. The actual noise seen in the output depends on how much bandwidth the sensor is allowed to pass.

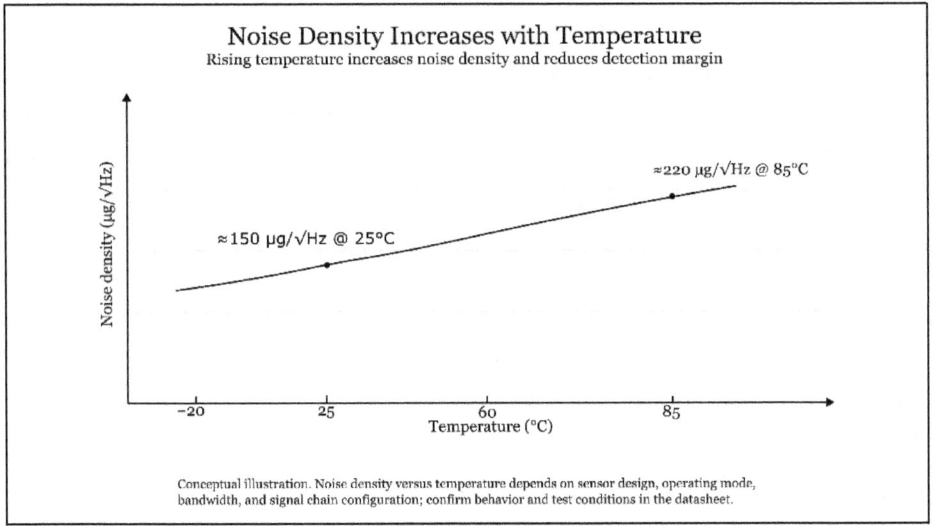

Figure 5.1 Noise density increasing with temperature.

As bandwidth increases, the accelerometer integrates noise over a wider frequency range. This means that even a sensor with low noise density can appear noisy if the bandwidth is set much higher than the application requires.

Using the relationship in Chapter 3:

$$RMS\ noise\ \approx\ noise\ density\ \times \sqrt{BW}$$

If an accelerometer has a noise density of $150\ \frac{\mu g}{\sqrt{Hz}}$ and the bandwidth (BW) is set to 100 Hz, the total noise is:

$$RMS\ noise\ \approx\ 150\ \frac{\mu g}{\sqrt{Hz}} \times \sqrt{100Hz}$$

$$RMS\ noise\ \approx\ 1500\ \ \mu g, or\ about\ 1.5\ mg$$

If the same sensor is configured for 400 Hz bandwidth, the total noise increases further, even though the hardware has not changed. Reducing the bandwidth to 50 Hz lowers the noise significantly, with no loss of useful information when the motion of interest is slow.

This explains why stationary accelerometers often show visible jitter and why increasing bandwidth tends to make the output look worse, not better. The noise is not a fault in the sensor. It is the expected result of allowing unnecessary frequency content through the signal path.

Most applications do not require high bandwidth. Step counting, posture monitoring, and sleep tracking usually need less than 25 Hz. Tilt sensing often works best below 10 Hz. Higher bandwidth is appropriate only when fast motion or high frequency content is truly present, such as in drones, robotics, or vibration analysis.

Misunderstanding the link between noise density, RMS noise, and bandwidth is one of the most common reasons engineers believe a sensor is too noisy, when the real issue is simply an overly aggressive bandwidth setting.

Noise does not exist in isolation. A common mistake is evaluating noise without considering how it interacts with full-scale range selection, filtering choices, and algorithm design. When these elements are treated separately, engineers often draw the wrong conclusions about sensor quality.

Selecting an unnecessarily large full-scale range reduces sensitivity and makes small real signals occupy fewer digital counts. The physical noise may remain the same in acceleration units, but the reduced counts per g make it harder to distinguish subtle motion from quantization limits and noise.

If a sensor provides $4096\frac{LSB}{g}$ in $\pm 2\ g$ mode and $1024\ \frac{LSB}{g}$ in $\pm 8\ g$ mode, the same physical RMS noise of 1 mg produces different digital jitter values. Because $1\ mg$ equals 0.001 g, the corresponding digital noise can be estimated using the sensitivity value.

In $\pm 2\ g$ mode:

$$0.001\ g \times 4096\frac{LSB}{g} \approx 4.1\ LSB$$

So the noise appears as about 4 LSB of jitter.

In $\pm 8\ g$ mode:

$$0.001\ g \times 1024\frac{LSB}{g} \approx 1.0\ LSB$$

So the same physical noise appears as about 1 LSB of jitter.

The sensor did not become quieter; the lower sensitivity in $\pm 8\ g$ mode simply maps the same physical noise into fewer digital counts.

Filtering is often misunderstood as a way to eliminate noise completely. In reality, filtering only reduces noise within a limited frequency range and always comes with tradeoffs. Aggressive filtering improves stability but slows response, adds latency, and distorts fast or transient motion.

Setting bandwidth to 5 Hz may produce smooth and stable tilt readings, but it also delays response and degrades gesture recognition or impact detection. Motion that occurs faster than the filter can pass is simply suppressed, even though it is physically present.

Noise must also be considered at the algorithm level. Thresholds placed too close to the noise floor are inherently unreliable. If RMS noise is about 1.5 mg, a detection threshold of 2 mg will occasionally trigger due to random fluctuations. This leads to false positives, missed events, and unstable behavior. Reliable algorithms require thresholds that sit comfortably above the noise floor and account for expected variation.

When noise is ignored in any one of these areas, range selection, filtering, or algorithm design, the system becomes fragile. Symptoms often appear as erratic step counts, unstable gesture detection, oscillatory control loops, or misleading vibration analysis. In most cases, the issue is not the sensor, but a failure to treat noise as a system-level constraint rather than a single parameter.

5.2.1 Bandwidth and ODR

One of the most persistent mistakes engineers make is assuming that ODR and bandwidth describe the same behavior. They do not. Although related, these parameters control different parts of the signal chain, and confusing them can lead to aliasing, unnecessary noise, poor responsiveness, or motion data that does not reflect what is actually happening.

Bandwidth controls which motion frequencies are allowed into the measurement path. It defines the highest frequency content that passes through the accelerometer's internal filtering before the signal reaches the ADC. If bandwidth is set to 100 Hz, motion above that frequency is attenuated or removed. Increasing bandwidth allows faster motion through, but it also allows more noise. Decreasing bandwidth suppresses high-frequency motion and reduces noise.

Figure 5.2 illustrates the magnitude response of a typical low-pass filter and how bandwidth is defined in practical sensing systems. Low frequency motion is preserved within the passband, where the response remains essentially flat. As frequency approaches the cutoff frequency f_C, the response drops to $-3\ dB$ relative to the passband, marking the onset of the transition region. Beyond this point, the filter increasingly attenuates higher-frequency content in the stopband, reducing noise and preventing distortion or aliasing in downstream signal processing. The gradual roll-off reflects the behavior of real filters, whose attenuation is finite and limited by filter order and implementation.

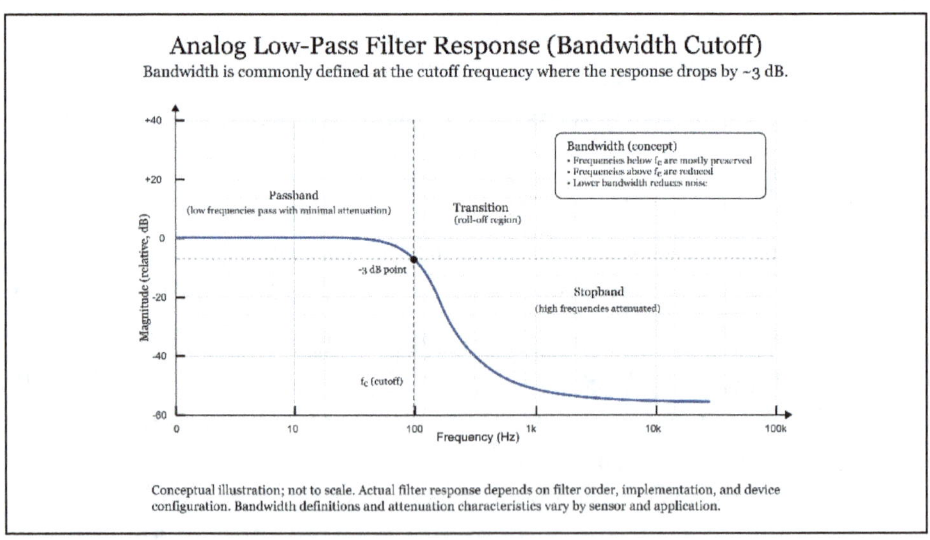

Figure 5.2 Analog low-pass filter response (bandwidth cutoff).

ODR controls how often the filtered signal is reported. If ODR is set to $100\ Hz$, the accelerometer delivers 100 samples per second to the host system. ODR affects time resolution, latency, and power consumption, but it does not by itself determine which motion frequencies are captured. Increasing ODR does not extend the measurable frequency range unless the bandwidth or filter settings are also changed; it only increases how often samples are produced.

Problems arise when bandwidth and ODR are not coordinated. If bandwidth is set higher than half the ODR, the sampled data no longer represents the input correctly. This violates the Nyquist requirement that ODR must be at least twice the bandwidth.

If bandwidth is set to 100 Hz and ODR is set to 50 Hz, motion above 25 Hz is misrepresented in the sampled data and may appear as false lower-frequency content. The resulting signal may still look reasonable, but it is physically incorrect.

Bandwidth selection can still be inadequate even when aliasing is avoided. Engineers sometimes raise bandwidth in an attempt to improve responsiveness, only to find that the signal becomes jittery and unstable. On the other hand, setting ODR far higher than necessary increases power consumption and processing load without improving signal quality. Setting ODR too low causes fast motion to be under-sampled, increasing latency and distorting waveforms.

The correct selection of bandwidth and ODR depends on the application and the highest meaningful motion frequency involved. For step counting, motion frequencies are typically between 1 Hz and 3 Hz. Bandwidths between 10 Hz and 25 Hz and ODR values between 25 Hz and 50 Hz are usually sufficient for reliable step detection while maintaining low power consumption.

For tilt sensing, bandwidths of 2 Hz to 5 Hz and ODR values of 10 Hz to 20 Hz provide stable readings with minimal jitter. For drone flight control, motion vibration may extend beyond 200 Hz. Bandwidth must be set near the highest meaningful vibration frequency, with ODR at least twice that value to avoid aliasing. For industrial vibration analysis, bandwidth and ODR must both be high enough to preserve frequency content, often several hundred Hz or more.

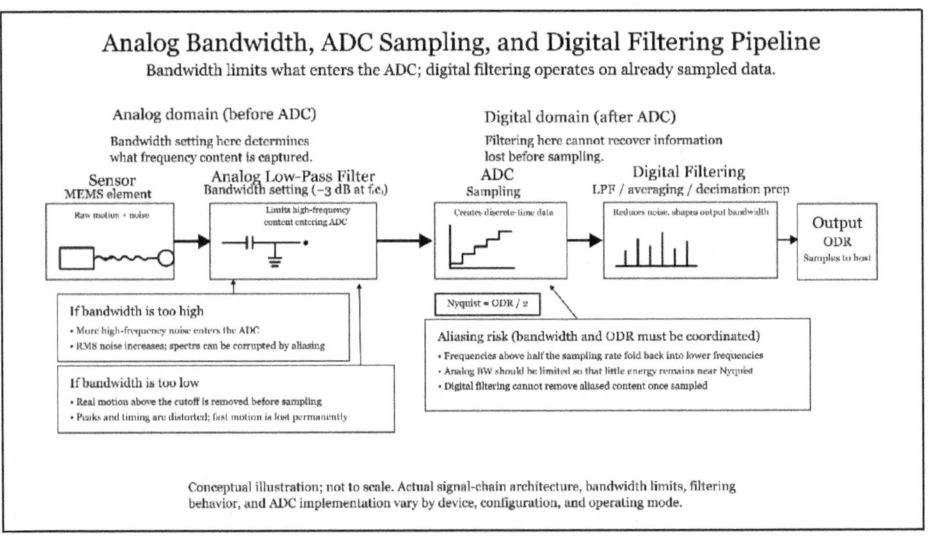

Figure 5.3 Analog bandwidth, ADC sampling, and digital filtering pipeline.

Figure 5.3 illustrates the signal path inside the accelerometer and reinforces an important practical point: the selected bandwidth determines which motion components reach the measurement chain, while later digital processing cannot recover information that was removed earlier.

Confusing bandwidth and ODR is dangerous because the resulting data often looks plausible even when it is fundamentally wrong. Understanding that bandwidth defines what motion is captured, while ODR defines how often it is reported, is essential for accurate measurement and stable system behavior.

As discussed in Section 5.2, increasing bandwidth also increases total RMS noise because a wider frequency range is admitted into the measurement path. The worked example in Section 5.2 shows how the same noise density produces a higher RMS noise value when bandwidth is increased. In practice, this means that selecting a bandwidth larger than the motion of interest can make slow or subtle signals appear noisier, even when the sensor itself is functioning correctly.

Setting bandwidth too low causes a different failure. Underselecting bandwidth suppresses real motion content, especially fast changes, making the system appear slow, unresponsive, or inaccurate. In practice, the selected bandwidth should be high enough to preserve the highest meaningful motion or vibration frequency in the application, but not so high that unnecessary noise is admitted.

Bandwidth and ODR should therefore be selected together. In addition, wider bandwidth usually increases power consumption because more analog and digital processing remain active. The most reliable approach is to choose the lowest bandwidth that still preserves the motion of interest while keeping the ODR high enough to sample it correctly.

5.3 Power Mode Misuse

Selecting a power mode that saves current but does not provide the bandwidth, ODR, latency, or internal processing needed by the application is a common design mistake. In practice, the differences between power modes can be substantial, and the wrong choice can cause missed events, delayed response, unstable outputs, or unnecessary power drain.

Figure 5.4 compares the performance characteristics of an accelerometer operating in active mode versus low power mode. Although low-power modes significantly reduce current consumption, they typically achieve this by lowering bandwidth, reducing internal sampling activity, limiting filtering options, and increasing noise. As a result, signal quality, responsiveness, and measurement fidelity can differ substantially from active mode.

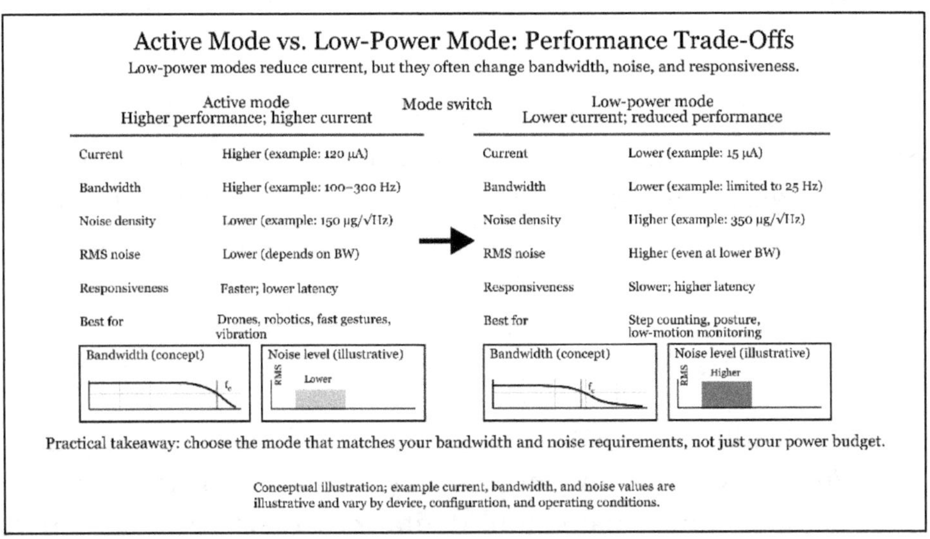

Figure 5.4 Active mode versus low-power mode performance trade-offs.

Active mode supports higher bandwidth, lower noise density, and faster response, making it suitable for applications such as vibration analysis, drones, robotics, and rapid motion tracking. Low-power mode is better suited for slow or intermittent motion, such as step counting or posture monitoring, where reduced bandwidth and higher noise are acceptable trade-offs.

This comparison highlights a common configuration mistake: assuming that low-power mode provides the same measurement performance as active mode while simply consuming less current. In practice, choosing the correct power mode is not only a power-budget decision, but also a signal-quality decision.

Power modes affect much more than current draw. They change how the analog front end operates, how filtering is applied, how often samples are generated, how interrupts behave, and sometimes even sensitivity itself. Systems that perform well in the lab often behave unpredictably in the field when these differences are ignored.

A frequent error is assuming that low power mode provides the same signal quality as active mode. This is rarely true. In low power modes, bandwidth is often reduced, internal sampling rates are lower, oversampling is limited, and noise density increases.

A datasheet may specify an active mode noise density of $150 \frac{\mu g}{\sqrt{Hz}}$ with bandwidth of $100\ Hz$, while low power mode specifies noise density of $350 \frac{\mu g}{\sqrt{Hz}}$ with bandwidth limited to $25\ Hz$. Although current consumption drops significantly, the sensor is no longer suitable for vibration analysis, fast gestures, drones, or robotics. The mode change alters what the sensor can reliably measure.

The reduced bandwidth does not itself cause the higher noise density. Instead, both changes result from the sensor entering a different operating mode. In low-power mode, the internal analog front end often runs at a lower-performance operating point to save current, which can raise the noise density while also limiting bandwidth. As a result, low-power operation may reduce power consumption but also degrade both frequency response and noise performance.

Wake-on-motion is another commonly misunderstood feature. It is designed for ultra-low power monitoring, not for accurate or continuous measurement. In this mode, only simplified detection logic is active. Data is not streamed continuously, thresholds are coarse, filtering is fixed or minimal, and detection latency is higher. Using wake-on-motion data to classify movement intensity or motion patterns produces unreliable results. Its purpose is to wake the system, not to replace active sensing.

Mode switching introduces its own pitfalls. When transitioning from standby or sleep to active mode, the analog circuitry and filters need time to stabilize. Reading data immediately after enabling active mode often produces invalid samples.

A datasheet may specify valid data availability 5 ms - 10 ms after enabling active mode. Reading data before this settling period can produce large offsets, excessive jitter, or false motion events that appear sporadic and difficult to debug.

Another mistake is overusing high-performance or high-resolution modes. These modes are intended for demanding applications that truly require higher bandwidth, lower noise, or faster response. Enabling them by default wastes power without improving results.

A wearable device running continuously in a high-performance mode may draw 120 μA instead of 15 μA in standard active mode. Battery life drops from months to weeks, even though the application only requires 10 Hz to 25 Hz motion tracking.

Noise behavior also changes across modes. Low power modes often exhibit significantly higher RMS noise due to reduced oversampling, simplified filtering, and low power clock sources.

If active mode RMS noise is 1.5 mg and low power mode RMS noise is 5 mg, an application attempting to detect 2 mg signals will fail in low power mode. The smallest meaningful signals are buried in noise, even though the same sensor performs well in active mode.

Using low power modes for high bandwidth applications creates predictable failure. If low power bandwidth is limited to 25 Hz, using this mode in a system that requires 200 Hz bandwidth causes severe signal attenuation, missed events, and unstable behavior. The sensor is operating as designed, but in the wrong mode.

Power modes are powerful tools when used intentionally. When misunderstood or misapplied, they lead to noisy data, missed motion, excessive power drain, or confusing behavior that is often blamed on the sensor rather than the configuration.

5.4 Orientation and Axis Alignment Errors

One of the most frequent and costly integration mistakes is misunderstanding how an accelerometer is oriented on the PCB. Accelerometers measure acceleration along their internal $X, Y,$ and Z axes as defined by the package, not by the PCB outline, enclosure shape, or product orientation. When these axes are misinterpreted or mapped incorrectly in software, motion data becomes misleading, even though the sensor itself is operating correctly.

As discussed in Chapters 3 and 4, axis misalignment means the sensor axes are not perfectly orthogonal. In practice, engineers usually encounter this not as an abstract specification, but as a system-level error that combines package orientation, PCB placement, coordinate-frame assumptions, and software interpretation mistakes.

Every accelerometer datasheet includes a diagram showing how the measurement axes are oriented. Errors occur because some diagrams show a top view while others show a bottom view, axis directions may be referenced to pin 1 or a package marking, and the *Z*-axis direction is not always intuitive.

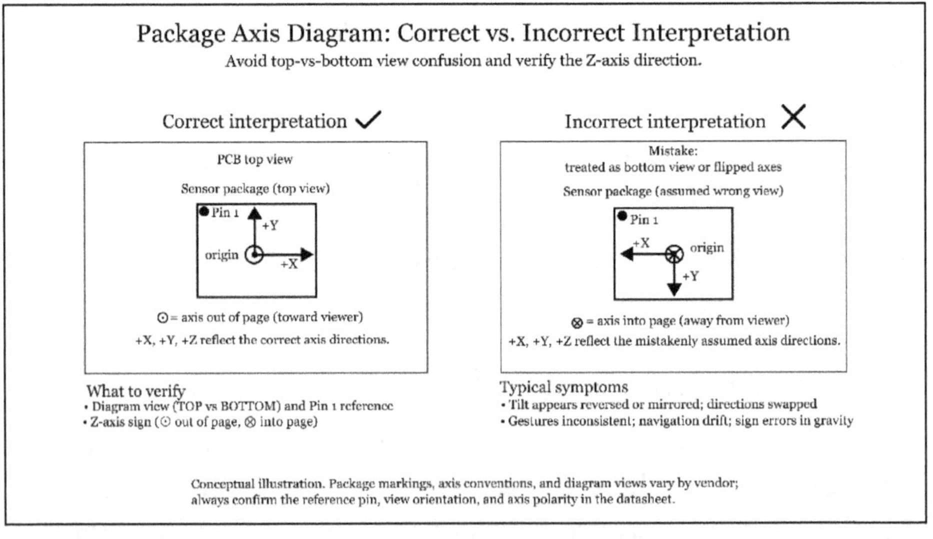

Figure 5.5 Package axis diagram: correct versus incorrect interpretation.

Figure 5.5 illustrates a common source of orientation errors: misinterpreting the package axis diagram in an accelerometer datasheet. The correct interpretation requires confirming whether the diagram represents a top or bottom view, identifying the Pin 1 reference, and verifying the direction of the Z-axis.

A frequent mistake is treating a top-view diagram as a bottom view or assuming the opposite Z-axis sign, which effectively mirrors or inverts the sensor axes. These errors do not indicate a faulty sensor; instead, they lead to reversed tilt, swapped directions, inconsistent gesture detection, or incorrect gravity interpretation. Careful reading of the package diagram and explicit verification of axis directions are essential before mapping sensor data in software.

For example, in a handheld device the system-level directions may be defined differently from the sensor's physical axes. Depending on how the package is mounted, the device's logical "up" direction may correspond to negative sensor Z, "forward" to positive sensor Y, and "left" to positive sensor X. The exact mapping depends on the product orientation and software coordinate-frame definition, which is why sensor axes must always be translated into the application's reference frame rather than assumed directly from package markings alone.

An engineer assumes that positive Z points away from the PCB, but the datasheet defines positive Z toward the PCB. Gravity is therefore interpreted with the wrong sign, producing a 180-degree error in tilt calculations. The sensor is correct, but the interpretation is not.

Figure 5.6 illustrates the difference between a system's logical coordinate frame and the accelerometer's physical sensor coordinate frame. Applications typically define motion in terms of system level axes such as forward, left, and up, which are chosen based on device orientation or user interaction rather than sensor placement.

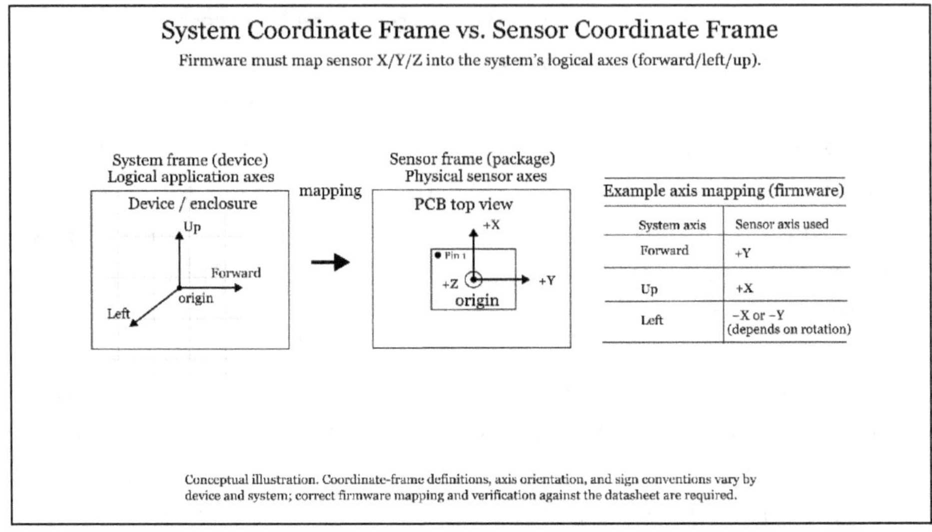

Figure 5.6 System coordinate frame versus sensor coordinate frame.

Axis errors also occur when a sensor is rotated on the PCB for routing or layout reasons and the software is not updated accordingly. If the sensor is mounted with a 90-degree rotation relative to the assumed orientation, forward motion may map to the wrong axis or appear mirrored. Without applying the proper coordinate transformation in software, all motion data is effectively rotated.

In contrast, an accelerometer reports measurements along its internal X, Y, and Z axes as defined by the package and mounting orientation on the PCB. When the sensor is rotated during layout, the sensor axes no longer align with the system axes, and the relationship must be explicitly handled in software through axis remapping, sign changes, or rotation matrices. Failing to apply this transformation causes motion to appear rotated, mirrored, or inconsistent across orientations, even though the sensor itself is operating correctly. Proper coordinate mapping ensures that sensor data is interpreted consistently and correctly within the application's reference frame.

Gravity further complicates axis interpretation. Engineers sometimes expect a stationary device to report zero on all axes, but this is incorrect. Gravity always contributes approximately $1\ g$ along one or more axes depending on orientation.

A device lying flat may report a_z close to $1\ g$, with a_x and a_y near zero. If the device is tilted slightly, gravity redistributes across multiple axes. This behavior is normal and does not indicate offset error or miscalibration. Misunderstanding gravity often leads engineers to suspect sensor drift when the readings are actually correct.

Even when axes are mapped correctly, accelerometer axes are not perfectly orthogonal. Datasheets typically specify axis misalignment on the order of a few degrees. A misalignment of 3° introduces a tilt error of about 5.2%, which can be significant in precision tilt sensing, robotics, or navigation. This estimate comes from basic trigonometry. If one sensing axis is misaligned by 3°, the unwanted component projected onto the wrong axis is approximately $sin(3°)\ \approx\ 0.052$ of the applied acceleration. For example, if $1\ g$ is applied along one axis, a 3° misalignment can produce an error of about $1\ g\ \times\ 0.052\ \approx\ 0.052\ g\ =\ 52\ mg$ on the other axis.

In real systems, orientation errors can cause drones to drift or oscillate, robots to move in unintended directions, wearables to misclassify activity, and industrial systems to misinterpret vibration direction. In many cases the sensor is functioning correctly, but the coordinate-frame interpretation is wrong.

5.5 Inadequate Filtering Choices

Filtering is essential when working with accelerometers, but inadequate filtering choices can distort signals, hide useful motion, add latency, or increase noise. Using the wrong filter type, selecting an inappropriate cutoff frequency, or applying filtering at the wrong stage can distort signals, introduce delay, or remove important motion content. A common mistake is assuming that more filtering automatically improves data quality. In practice, filtering is always a tradeoff between noise reduction and responsiveness. Filtering choices affect every downstream process, including gesture recognition, vibration analysis, control loops, and power consumption. When filtering is inadequately chosen, systems often feel sluggish, unstable, or misleading, even though the sensor itself is operating correctly.

Figure 5.7 illustrates where filtering occurs in the accelerometer signal chain and why filtering choices must be made deliberately. Analog low-pass filtering operates before the ADC and determines which frequency content enters the sampling process. This stage is critical because errors introduced here cannot be fully corrected later by digital filtering.

Digital filtering occurs after sampling and can smooth already sampled data, reduce noise, and shape the output bandwidth, but it cannot undo aliasing or recover motion that was never captured. The figure therefore highlights a common misconception: digital filtering cannot compensate for incorrect analog bandwidth selection. Effective filtering requires coordinating analog bandwidth, ODR, and digital filtering so that noise, latency, and signal fidelity are balanced for the intended application.

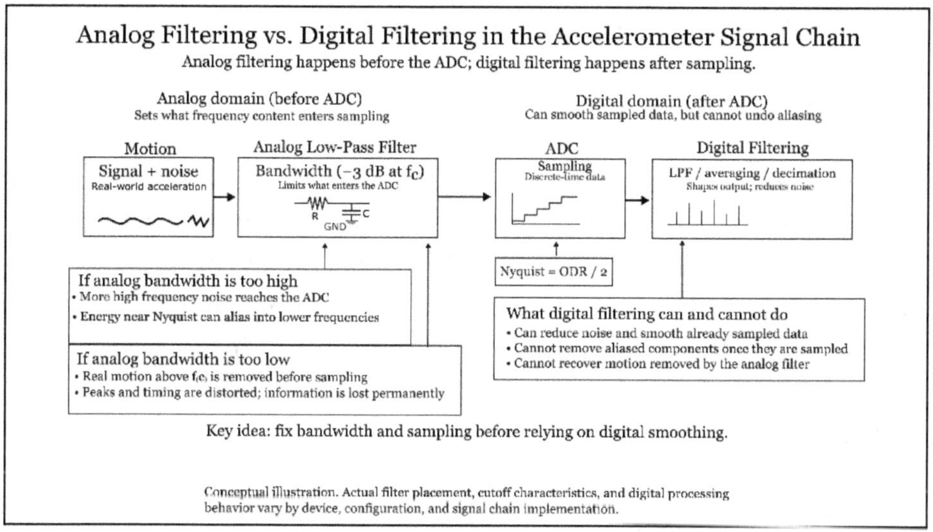

Figure 5.7 Analog and digital filtering in the accelerometer signal chain.

Setting the cutoff frequency too low is a frequent error. When the cutoff is far below the actual motion content, the signal becomes overly smooth and delayed. A gesture-recognition system using a cutoff frequency of 3 Hz suppresses rapid hand motion. Gesture peaks are attenuated, response is delayed, and classification becomes unreliable. Step impacts may disappear entirely, making the system feel unresponsive. The opposite mistake is setting the cutoff frequency too high. Allowing unnecessary bandwidth through the filter increases noise without revealing useful information.

Filtering a signal that only requires 20 Hz bandwidth with a 200 Hz cutoff increases RMS noise ($RMS\ noise \propto Bandwidth$) by roughly

$$\sqrt{\frac{200}{20}} = \sqrt{10}$$

The result is increased jitter, reduced signal-to-noise ratio, and unstable outputs, especially in tilt sensing, posture analysis, and sleep monitoring.

Another common misunderstanding is assuming that digital filtering alone can compensate for improper analog bandwidth selection. If analog bandwidth is too high, noise and aliased components enter the ADC and cannot be removed later. If analog bandwidth is too low, high frequency motion is lost permanently. Digital filters cannot recover information that was never captured.

Overreliance on simple moving average filters is another frequent mistake. Moving averages are easy to implement but introduce significant delay and distort transient motion. A ten-sample moving average at an ODR of 100 Hz introduces about 100 ms of delay. This may be acceptable for slow activity tracking, but it is unsuitable for drones, robotics, gaming, or gesture recognition, where responsiveness matters.

The delay comes from the fact that a moving average uses multiple consecutive samples to produce each output value. At an ODR of 100 Hz, one new sample arrives every 10 ms. A ten-sample moving average therefore spans $10 \times 10\,ms = 100\,ms$ of data. In practice, the effective delay is tied to that averaging window, because the output cannot fully reflect a sudden change until enough new samples have entered the average. Larger moving-average windows increase smoothing, but they also increase response delay and blur fast transients.

All filters introduce latency, and aggressive filtering makes this delay more noticeable. In wearables, some delay may be acceptable. In control systems, the same delay can cause oscillation or instability. Smoothness and responsiveness must be balanced intentionally.

Filtering raw acceleration without accounting for gravity also causes problems. Raw acceleration includes both dynamic motion and gravity. Filtering them together distorts both components. For dynamic motion analysis, gravity should be removed first:

$$a_{dynamic} = raw_{accel} - g_{vector}$$

In practice, the gravity component is typically estimated using a low-pass filter applied to the raw acceleration signal over time. This filter tracks the slowly changing component of the signal, which corresponds to gravity and slow orientation changes. The estimated gravity vector is then subtracted from the raw measurement to obtain the dynamic acceleration. Any further filtering should then be applied to the gravity-compensated signal rather than to the original combined measurement.

High-pass filters are often misused for this purpose. A high-pass filter removes low-frequency content, including gravity. When a high-pass filter is enabled without understanding its effect, static orientation information becomes meaningless. As a result, tilt estimation and posture tracking fail, even though dynamic motion may still appear valid. High-pass filtering is appropriate only when gravity and long-term orientation are irrelevant, such as in impact detection or vibration analysis.

Filter behavior can change across power modes. Low power modes often use simplified or fixed filters that differ from active mode filtering. A low power mode may enforce a fixed 10 Hz low-pass filter. Even if the ODR appears adequate, fast gestures become unreliable because the filter suppresses the motion before it reaches the output.

Ineffective filtering choices rarely cause obvious failure. Instead, they produce systems that feel noisy, slow, unstable, or inconsistent. In many cases, the sensor is blamed when the real issue is simply a filtering strategy that does not match the motion being measured.

5.6 Failing to Prevent Aliasing

Aliasing is one of the most damaging signal-processing mistakes engineers make when working with accelerometers. It occurs when high-frequency motion is sampled too slowly, causing that motion to appear as false low-frequency signals. Because the resulting data often looks plausible, aliasing can go unnoticed and lead to incorrect interpretation of motion or vibration behavior.

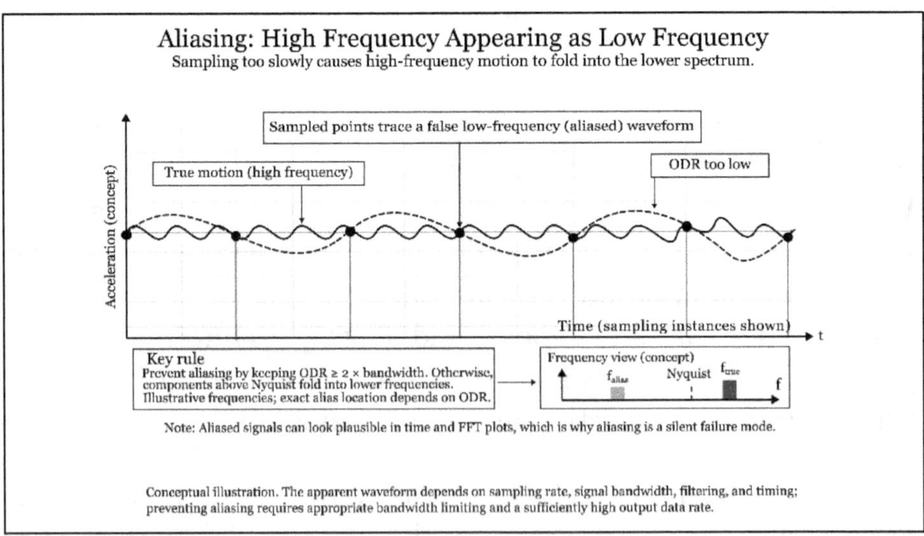

Figure 5.8 Aliasing from undersampling: high-frequency motion appearing as a false low frequency.

Figure 5.8 illustrates how aliasing arises when an accelerometer samples motion at a rate that is too low relative to the signal bandwidth. The solid waveform represents the true high-frequency physical motion, while the sampled points represent the data captured by the ADC. Because the ODR is insufficient, the sampled points appear to follow a different, lower-frequency pattern than the original signal.

In this context, ODR represents the effective output sampling rate, meaning how often new acceleration data is made available at the sensor output. If that rate is too low relative to the motion bandwidth, aliasing can occur even if the reported waveform appears plausible.

This effect occurs whenever the signal frequency exceeds half the sampling rate, known as the Nyquist frequency. When this limit is exceeded, high-frequency motion cannot be represented correctly and instead appears as lower-frequency components in the sampled data.

When a signal is sampled too slowly, successive samples no longer capture each cycle of high-frequency motion. Instead, the sampled points align in a pattern that mimics a lower-frequency waveform, producing misleading measurements and incorrect spectral interpretation.

Aliasing is especially problematic because the data often looks realistic even when it is fundamentally wrong. For vibration analysis, this can produce plausible but incorrect periodic motion information that cannot be corrected later. As explained in Section 3.2, preventing aliasing requires that the ODR remain high enough relative to the analog signal bandwidth. The basic requirement is simple:

$$ODR \geq 2 \times bandwidth$$

If this condition is violated, frequency components above half the ODR appear in the lower spectrum. If accelerometer bandwidth is set to 200 Hz, the ODR must be at least 400 Hz. If ODR is instead set to 100 Hz, all motion above 50 Hz aliases into lower frequencies, producing incorrect motion information that may still appear plausible.

In most accelerometers with digital output, bandwidth is not configured directly. Instead, the effective signal bandwidth is determined by the selected ODR and the sensor's internal filtering, with datasheets typically specifying the resulting bandwidth for each ODR setting.

Aliasing occurs because accelerometers respond to all mechanical motion within their bandwidth, including vibration, shocks, and structural resonance. If the analog bandwidth is wider than what the system can sample, high-frequency components pass through the analog front end and are misrepresented by the ADC. Preventing aliasing therefore depends primarily on proper bandwidth selection before digitization.

The consequences of aliasing are most visible in vibration and control applications. An industrial motor produces vibration at 300 Hz, but the accelerometer is configured with an ODR of 100 Hz. The 300 Hz vibration aliases into a much lower apparent frequency, often around 20 Hz or 30 Hz. Frequency analysis becomes misleading, bearing faults go undetected, and diagnostic conclusions are incorrect.

In drones, rotor vibration near 250 Hz can alias into the control bandwidth if sampling is too slow. The flight controller interprets the aliased signal as drift or oscillation, leading to unstable behavior even though the hardware is functioning properly.

A common misconception is that digital filtering can remove aliasing artifacts. This is incorrect. The signal path is:

$$Mechanical\ motion \rightarrow analog\ filtering \rightarrow ADC \rightarrow digital\ filtering$$

If high frequency motion enters the ADC because bandwidth is too high or sampling is too slow, the ADC produces distorted samples. Digital filters can smooth those samples, but they cannot reconstruct the original signal. Once aliasing occurs, the original frequency content is permanently lost.

The aliased frequency is the false lower-frequency component that appears in the sampled output when the true signal is sampled too slowly. Estimating this apparent frequency helps engineers recognize when a suspicious low-frequency oscillation may not be real, but instead the result of undersampling. This is especially useful in vibration analysis and control systems, where aliasing can produce misleading frequency content that appears physically plausible.

Given the true signal frequency and the sampling frequency, the aliased frequency can be estimated as:

$$f_{alias} = | f_{signal} - N \cdot f_s |,$$

where f_{alias}, f_{signal}, and f_s are the aliased frequency, true signal frequency, and sampling frequency, respectively, and N is the nearest integer multiple.

For example, if the true vibration frequency is 180 Hz and the ODR is 100 Hz, the aliased frequency can be computed as:

$$f_{alias} = | 180 - 2 \times 100 | = 20\ Hz.$$

Here, $N = 2$ because $2f_s = 200\ Hz$ is the multiple of the sampling rate closest to the true signal frequency of 180 Hz.

The sensor therefore reports a slow 20 Hz oscillation even though the physical vibration is much faster. This false frequency is convincing and dangerous if not recognized.

Aliasing often reveals itself indirectly. Engineers may observe unexpected low frequency oscillations, unstable control behavior, inconsistent FFT (Fast Fourier Transform) peaks, or vibration patterns that do not match physical intuition. When frequency domain results appear suspiciously low or inconsistent, aliasing should always be considered.

Aliasing is a silent failure mode. Once understood and prevented through proper coordination of bandwidth and ODR, system behavior becomes far more predictable and reliable.

5.6.1 Failing to Calibrate the Accelerometer

A surprisingly common mistake, even among experienced engineers, is assuming that an accelerometer will behave correctly without calibration. Although modern MEMS accelerometers are factory trimmed and generally consistent, they are not perfectly calibrated for every application, mounting condition, or operating environment. Ignoring calibration leads to offset errors, incorrect tilt estimates, device-to-device variation, and unstable motion interpretation.

Calibration is not an optional refinement. It is a fundamental step for achieving reliable and repeatable performance, especially in applications involving tilt, posture, vibration analysis, or multi-axis motion.

Factory calibration corrects average offset, sensitivity, and temperature behavior under ideal conditions. Real systems rarely match those conditions. PCB stress, enclosure pressure, solder reflow, mounting orientation, and local heating all shift sensor behavior after assembly.

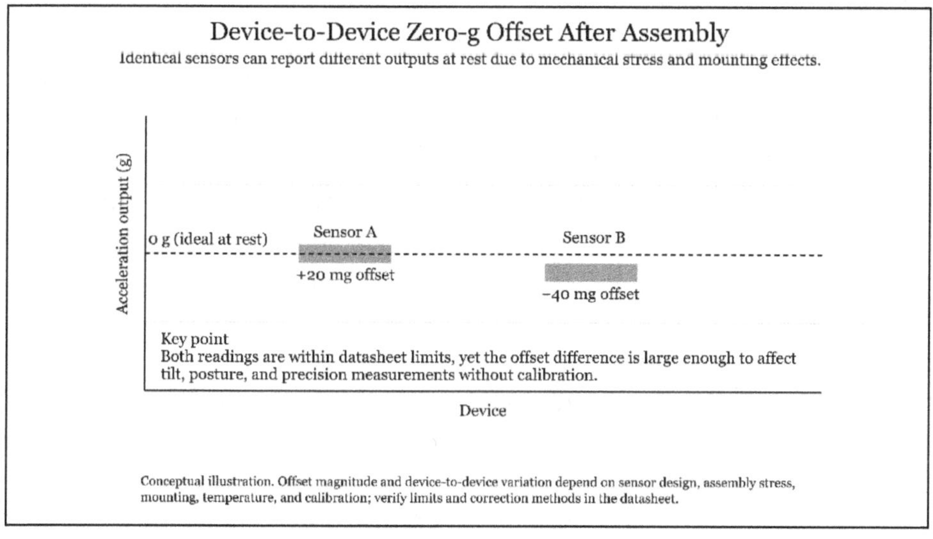

Figure 5.9 Device-to-device zero-g offset after assembly.

Figure 5.9 illustrates how two nominally identical accelerometers can report different outputs when no acceleration is present. Although both sensors are stationary and should ideally measure 0 g, mechanical stress from PCB mounting, solder reflow, enclosure constraints, or local heating introduces offset shifts after assembly.

In this example, Sensor A exhibits a $+20\ mg$ offset while Sensor B shows a $-40\ mg$ offset. Both values may fall within datasheet specifications, yet the difference between them is large enough to affect tilt estimation, posture tracking, and other precision measurements. Without offset calibration, these device-to-device variations lead to inconsistent behavior across units and degrade accuracy even in static conditions.

Offset calibration addresses this problem by correcting nonzero output when an axis should measure zero acceleration. A simple approach is to place the device motionless and record repeated readings from the same unit. A simple arithmetic mean over tens to hundreds of samples is typically sufficient, because averaging reduces random noise and provides a more reliable estimate of the underlying DC offset. Very noisy systems or measurements taken in the presence of vibration may require more careful data collection. For the X and Y axes, the average value directly represents the offset. For the Z axis, gravity must be accounted for.

If the device is flat and stationary, offsets can be computed as:

$$x_{offset} = average(a_x)$$

$$y_{offset} = average(a_y)$$

$$z_{offset} = average(a_z - 1\ g)$$

These offsets are then subtracted from all future measurements. Skipping this step leaves static error uncorrected and degrades tilt and posture accuracy.

Sensitivity calibration is often overlooked but equally important. Datasheets commonly specify sensitivity tolerance of several percent, meaning two sensors configured identically may produce different digital values for the same physical acceleration.

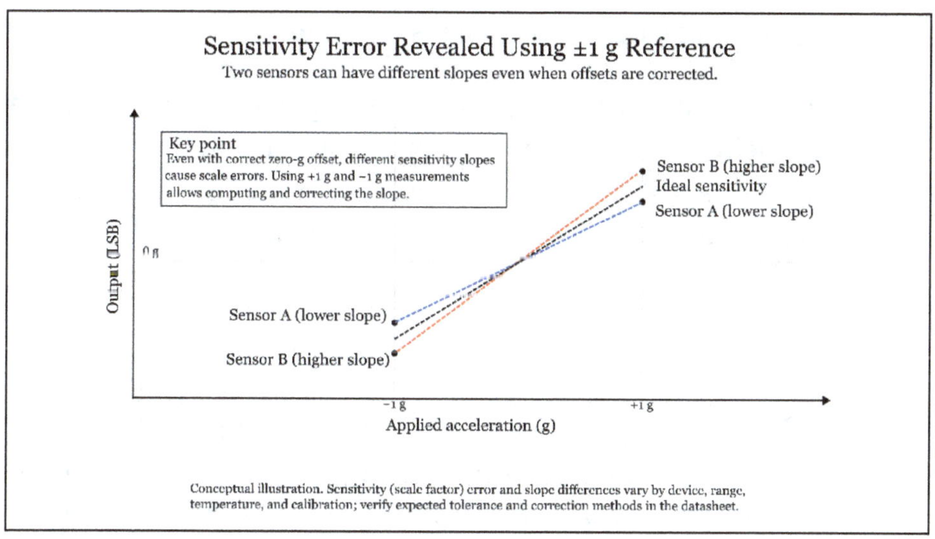

Figure 5.10 Sensitivity error revealed using ±1 g reference.

Figure 5.10 illustrates sensitivity, or scale factor, error in an accelerometer after zero-g offset has already been corrected. The dashed reference line represents the ideal sensitivity specified in the datasheet, while the other two lines show the measured responses of Sensor A and Sensor B. Although both sensors read zero correctly at $0\ g$, their outputs at $-1\ g$ and $+1\ g$ differ, resulting in different slopes. This slope difference means that the same physical acceleration produces different digital values, leading to scale errors in tilt estimation, motion magnitude, and vibration analysis. Sensitivity calibration uses the known $\pm 1\ g$ gravity reference to compute the correct slope and apply a scale factor in software, ensuring consistent and accurate measurements across devices.

If nominal sensitivity is $1024\frac{LSB}{g}$ with a $\pm 5\%$ tolerance, one device may produce $972\frac{LSB}{g}$ while another produces $1075\frac{LSB}{g}$. Over a $\pm 1\ g$ range, this difference is significant.

A simple sensitivity calibration uses gravity as a reference. Rotating the device so an axis experiences $+1\ g$ and $-1\ g$ yields two measurements. This method is valid for any selected full-scale range that accommodates $\pm 1\ g$. Sensitivity is then computed as:

$$sensitivity = (reading_{high} - reading_{low}) \ / \ 2\ g$$

The denominator is $2\ g$ because the input changes from $-1\ g$ to $+1\ g$, corresponding to a total acceleration span of $2\ g$. Applying this scale factor greatly improves accuracy and consistency.

Axis alignment can also require calibration when the sensor axes are not perfectly orthogonal. As discussed in Section 5.5 even a few degrees of misalignment can introduce measurable cross-axis error, so precision applications may require a correction matrix based on known reference orientations.

Temperature effects further complicate calibration. Offset and sensitivity vary with temperature, and calibrating only at room temperature assumes behavior remains constant, which is rarely true. If offset drift is $1\frac{mg}{°C}$, a $40°C$ temperature change introduces $40\ mg$ of error. For low frequency or tilt-based applications, this error is substantial.

A simple temperature compensation model stores calibration parameters at multiple temperatures and interpolates between them, for example:

$$offset(T) = offset_{25} + k_{offset} \times (T - 25°C)$$

Where:

$offset(T)$ is the estimated sensor offset at temperature T,

$Offset_{25}$ is the measured offset at the reference temperature of 25 °C,

k_{offset} is the offset temperature coefficient (change in offset per degree Celsius),

Using the internal temperature sensor makes this approach practical. Calibration quality also depends on conditions during calibration. Performing calibration while the device is vibrating, being handled, or resting on a flexible surface produces incorrect results. Calibrating on a desk that vibrates slightly due to nearby equipment introduces bias that later appears as unexplained drift. Calibration must be performed when the device is completely still and mechanically stable.

Using two or three calibration points dramatically improves stability over temperature. Temperature effects also influence algorithm behavior. Fixed thresholds that work at one temperature often fail at another.

Figure 5.11 illustrates how offset error varies with temperature and how multi-point calibration improves stability. The raw temperature-dependent offset curve is characterized at multiple calibration temperatures, and compensation parameters are interpolated between those points. Compared to single-point calibration, this approach significantly reduces residual error across the operating range and improves consistency in environments with large temperature variation.

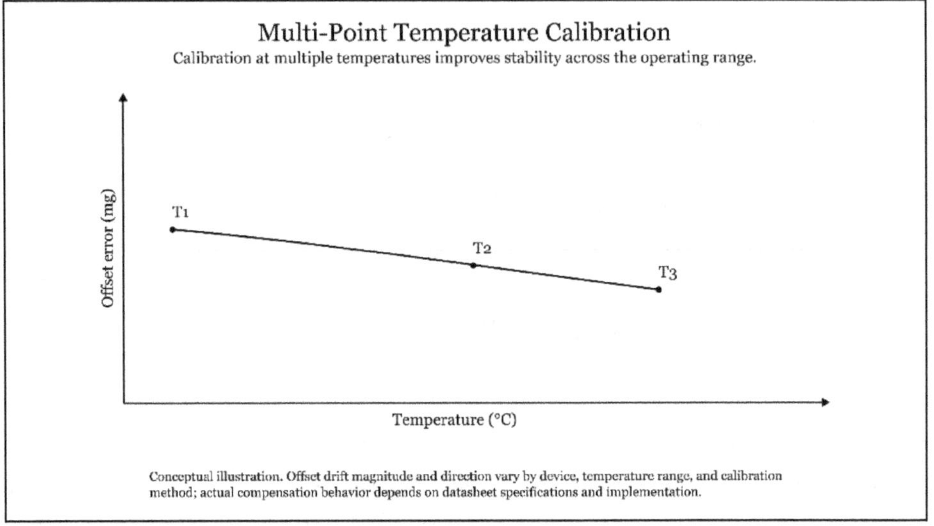

Figure 5.11 Multi-point temperature calibration curve.

In practice, offset variation with temperature is not perfectly linear, so using multiple calibration points allows the compensation model to better capture this curvature and reduce residual error across the operating range.

Failing to calibrate does not usually cause immediate or obvious failure. Instead, it produces systems that behave inconsistently, drift over time, or vary from unit to unit. In many cases, the sensor is blamed when the real issue is simply missing or incomplete calibration.

Calibration results are not transferable across mechanical changes. Changes in PCB layout, enclosure design, screw torque, adhesive use, or even production batch can alter sensor stress and invalidate previous calibration. Reusing old calibration values after such changes leads to degraded performance that is often misdiagnosed as sensor failure.

5.7 Sensor Placement and Mechanical Stress

Even with correct configuration, calibration, and filtering, poor physical placement or mechanical stress can undermine accelerometer performance in subtle and persistent ways. MEMS accelerometers contain microscopic mechanical structures that respond not only to external acceleration, but also to stress in the package and silicon.

PCB flexing, enclosure pressure, thermal expansion, and mounting choices can shift offset, alter sensitivity, and introduce cross-axis effects that vary over time and temperature. This mistake is particularly destructive because mechanical effects often masquerade as electrical noise, temperature drift, or algorithm instability. Engineers may spend weeks tuning filters or rewriting code while the real problem is mechanical.

Because these stress effects are highly location-dependent, sensor placement on the PCB matters greatly. Regions near mounting screws, board edges, cutouts, thin sections, or local heat sources are more likely to experience flex, strain, or thermal gradients that shift the measured output. These placement-dependent stresses can produce apparent drift, offset shifts, or cross-axis errors that filtering and calibration cannot fully remove. Proper placement is therefore essential for stable and repeatable accelerometer performance.

Figure 5.12 compares recommended and less suitable accelerometer placement on a PCB, showing how placement near flex zones, screws, cutouts, or heat sources can expose the sensor to mechanical stress and thermal gradients that may degrade measurement accuracy. In the recommended case, the sensor is placed in a rigid region away from edges, corners, mounting features, and local heat sources. In the less suitable case, the sensor is located near a cutout, screw-stress region, flex zone, and hot component, where board deformation and local thermal variation are more likely to influence the measured output.

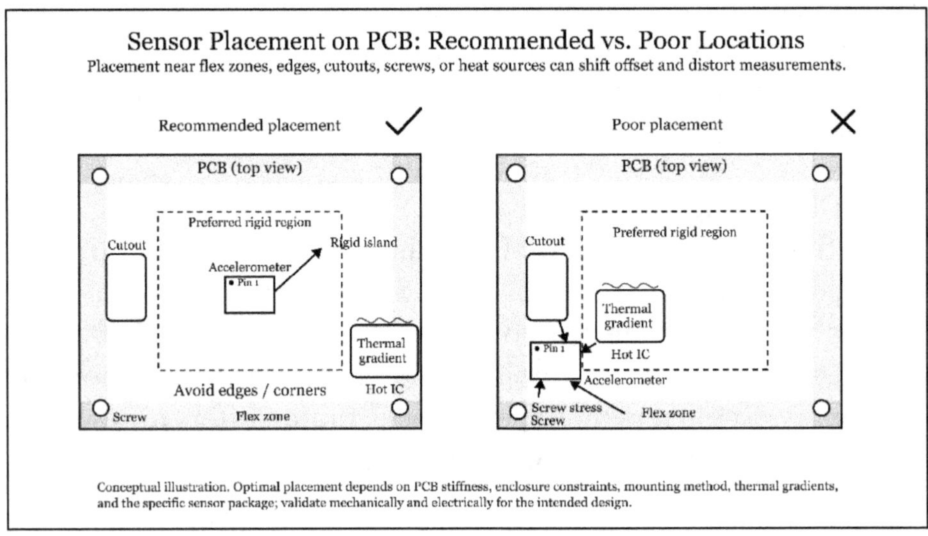

Figure 5.12 Recommended versus less suitable accelerometer placement on a PCB.

Enclosure design is another common source of stress. Pressure from plastic snaps, over-tightened screws, rubber pads, or warped housings can compress the accelerometer package. Even small forces distort the MEMS structure and shift the zero-g offset.

For example, a device may report $a_x = +0.015\ g$ before enclosure assembly and $a_x = +0.065\ g$ afterward. This 50 mg shift is often blamed on temperature effects or sensor quality, even though the true cause is enclosure-induced stress. Because these effects vary from unit to unit, they can also create device-to-device inconsistency.

Thermal stress can produce similar symptoms. Materials on a PCB expand at different rates as temperature changes. If an accelerometer is placed near heat sources such as voltage regulators, processors, power transistors, or radios, uneven expansion bends the PCB locally. As a nearby regulator warms during operation, the PCB bows slightly, shifting the accelerometer offset over several minutes. The drift therefore follows how long the system has been powered and how much nearby components have heated, rather than the surrounding air temperature alone. This makes diagnosis especially confusing, because the behavior may look like a sensor or temperature-drift problem even though the real cause is local board deformation from uneven heating.

In wearables and compact devices, flexible or semi-flexible PCBs are common. Mounting an accelerometer directly on a flexible section is a serious mistake. As the PCB bends during normal use, offsets vary with posture, motion, or strap tension, making tilt and posture data unreliable. Whenever possible, accelerometers should be placed on rigid PCB islands, even within flexible assemblies.

Beyond static stress and thermal gradients, mechanical coupling can also degrade accelerometer measurements. Vibration from motors, fans, pumps, high-airflow systems, or other moving structures can travel through the PCB or enclosure and reach the sensor. If these vibrations align with the accelerometer's mechanical resonance or fall near a frequency of interest, the measured output can become amplified or distorted.

Figure 5.13 illustrates how mechanical vibration generated by components such as motors can propagate through the enclosure and PCB and influence accelerometer measurements. Vibrations travel along structural paths and can couple directly into the sensor if it is placed near the excitation source or on a mechanically compliant region of the board.

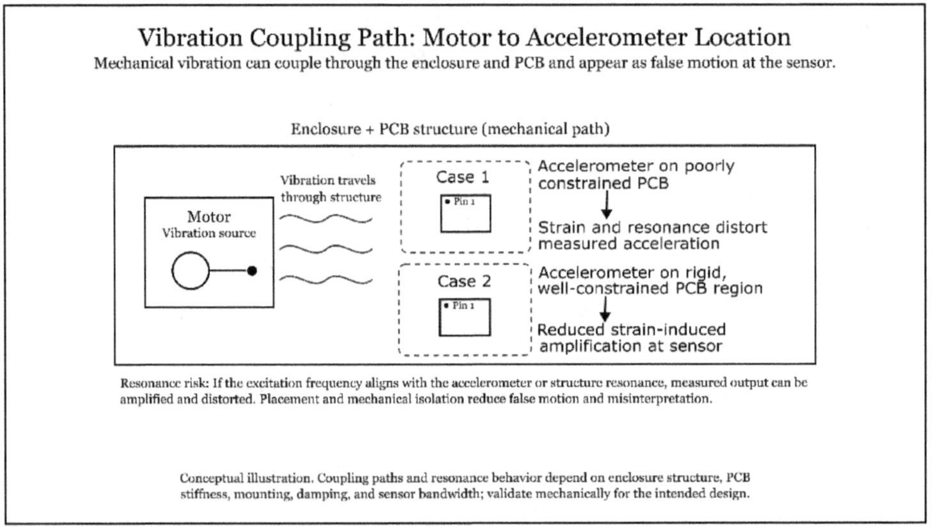

Figure 5.13 Mechanical vibration coupling from motor to accelerometer location.

In such cases, the accelerometer reports amplified or distorted signals that may be misinterpreted as real motion. Placing the sensor on a rigid PCB island, increasing mechanical separation from vibration sources, and adding damping or isolation reduce the amount of vibration reaching the sensor.

Proper placement and mechanical design are therefore essential to prevent resonance, false motion detection, and unstable behavior in vibration-sensitive applications. For example, a vibration motor operating near 175 Hz may induce exaggerated readings on one axis, confusing gesture detection or activity classification algorithms even though the sensor itself is functioning correctly.

Adhesives and underfill can also introduce long-term mechanical stress. If applied unevenly or excessively, they pull on the package as they cure or age, permanently shifting offset or sensitivity. These effects often appear only after environmental testing or extended field use, making them difficult to diagnose.

Mechanical stress rarely causes obvious failure. Instead, it produces systems that drift, behave inconsistently between units, or change over time. The real issue is often mechanical integration, especially where and how the sensor is mounted.

5.8 Misinterpreting Motion Patterns

Motion data from accelerometers is rich but complex. A common mistake is assuming that raw acceleration signals directly reveal steps, falls, gestures, or vibration behavior in a simple and consistent way. In reality, accelerometer data is influenced by gravity, orientation, filtering, bandwidth, noise, and timing. Misinterpreting motion patterns leads to false detections, unstable algorithms, and incorrect conclusions about system behavior.

Correct motion interpretation requires understanding how physical events unfold over time, not how they appear in isolated samples. One frequent error is assuming that motion patterns are independent of orientation. A step pattern measured with the device upright may show clear peaks on one axis. Rotating the device by 90° shifts those peaks to a different axis and changes their amplitude. Algorithms that assume a fixed axis fail as soon as orientation changes.

Figure 5.14 shows a representative acceleration waveform for a single step, highlighting the distinct phases that occur over time. A complete step cycle includes upward acceleration, downward motion, a brief impact peak at foot contact, and a recovery phase before the next step begins. These features vary in amplitude and timing across users, walking speeds, footwear, and device placement.

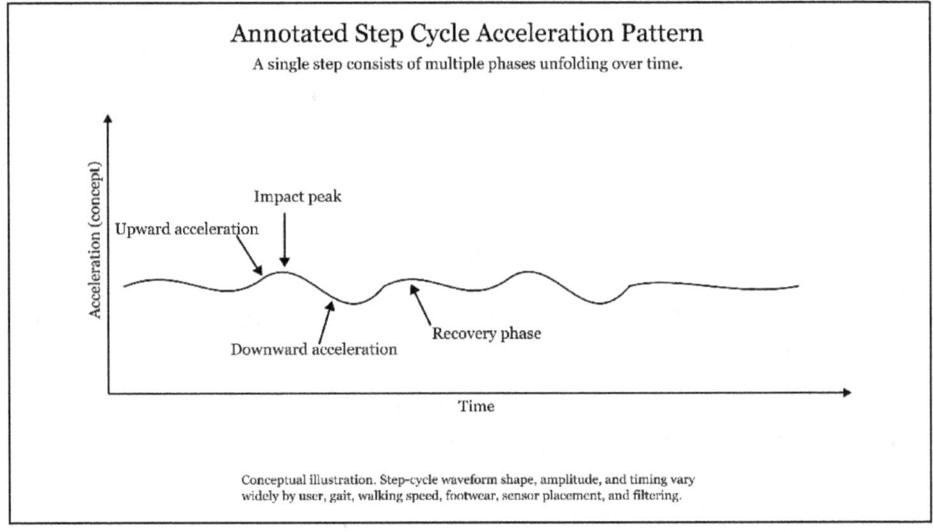

Figure 5.14 Annotated step cycle acceleration pattern.

Reliable step detection therefore depends on recognizing the temporal structure (timing pattern) of the entire step cycle rather than reacting to isolated peaks or threshold crossings.

Beyond the example shown in **Figure 5.14**, real step signatures vary widely across users, walking speeds, footwear, device placement, and filtering conditions. Excessive filtering can suppress impact peaks, while insufficient filtering allows noise to trigger false steps. Reliable step detection therefore depends on recognizing the expected step sequence while tolerating normal variation in amplitude, timing, and waveform shape.

Gravity itself is often confused with dynamic motion. Gravity contributes to a constant acceleration of approximately 1 g, and changes in orientation redistribute this component across axes. Tilting a stationary device causes a_z to decrease while a_x or a_y increases. This is not dynamic movement, but a change in orientation relative to gravity. Treating this as motion leads to false detections and unstable baselines. Separating gravity from dynamic acceleration is essential for reliable motion interpretation.

Falls are frequently misclassified for similar reasons. Engineers often rely on peak acceleration alone, which is not sufficient. A rapid arm movement may exceed a high acceleration threshold but does not represent a fall. A true fall includes a sequence of near-zero acceleration, a high impact peak, and subsequent stillness. Detecting only the peak produces frequent false alarms.

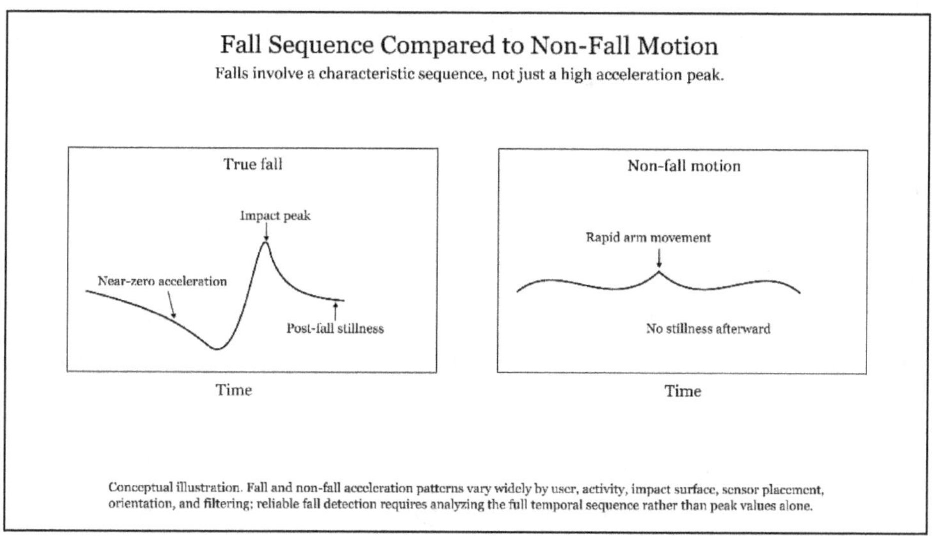

Figure 5.15 Fall sequence compared to non-fall motion.

Figure 5.15 compares the acceleration patterns of a true fall and a non-fall motion such as a rapid arm movement. A fall typically includes a brief period of near-zero acceleration during free fall, followed by a high impact peak and subsequent stillness as the body comes to rest. In contrast, non-fall motions may exhibit high acceleration peaks but lack the characteristic sequence and post event stillness. Reliable fall detection therefore depends on recognizing the full temporal pattern rather than relying on peak acceleration alone.

Vibration signals present a different challenge because their interpretation is especially sensitive to configuration. Bandwidth, ODR, and aliasing strongly influence how vibration appears in both time and frequency domains. A vibration spectrum showing a dominant $30\,Hz$ component may actually represent much higher physical vibration that has aliased due to insufficient sampling. Without proper configuration, frequency-domain analysis becomes misleading. Mechanical coupling, resonance, and PCB placement further complicate interpretation.

Gestures are another source of false assumptions. Engineers often expect gestures to be clean and repeatable. Two users performing the same gesture may produce acceleration patterns with very different amplitudes, durations, and axis contributions. Expecting consistent waveforms leads to fragile detection logic. Robust gesture recognition relies on timing, pattern recognition, and multi-axis features rather than exact waveform matching.

Many motion events involve all three axes simultaneously. Analyzing only one axis often produces incomplete or misleading results. A hand motion may combine vertical movement, lateral sweep, and rotation. Observing only one axis hides important information and weakens classification accuracy. Using vector magnitude or cross-axis analysis improves robustness across orientations and users.

Misinterpreting motion patterns does not usually cause an obvious failure. Instead, it leads to systems whose behavior changes across users, orientations, and environments. In most cases, the real problem is an oversimplified interpretation of motion that is more complex than a single waveform or threshold can represent.

5.9 Misusing Thresholds and Interrupts

Modern digital accelerometers include powerful interrupt features such as wake-on-motion, activity detection, freefall detection, tap and double-tap detection, and programmable motion thresholds. When used correctly, these features reduce power consumption and simplify system design. A common mistake is treating interrupts as simple comparisons against raw acceleration data.

In reality, interrupt engines often operate on separate signal paths with their own filtering, timing, and internal sampling behavior. Misunderstanding these differences leads to missed events, excessive false triggers, and system behavior that appears inconsistent or unpredictable.

One frequent error is setting thresholds too close to the noise floor in an attempt to maximize sensitivity. If RMS noise is about 1.5 mg and a motion threshold is set at 2 mg, random noise fluctuations will occasionally exceed the threshold. The result is repeated false interrupts even when the device is motionless. Reliable detection requires thresholds that sit comfortably above the noise floor, not just barely above it.

Figure 5.16 illustrates how setting an interrupt threshold too close to the noise floor leads to repeated false triggers. Even when the device is motionless, random noise fluctuations cause the acceleration signal to jitter above and below the threshold, generating interrupts that do not correspond to real motion. Reliable interrupt-based detection requires thresholds that are comfortably above the RMS noise level, along with appropriate filtering, hysteresis, or debounce logic (brief trigger suppression). Thresholds chosen solely to maximize sensitivity often produce unstable system behavior and excessive false events.

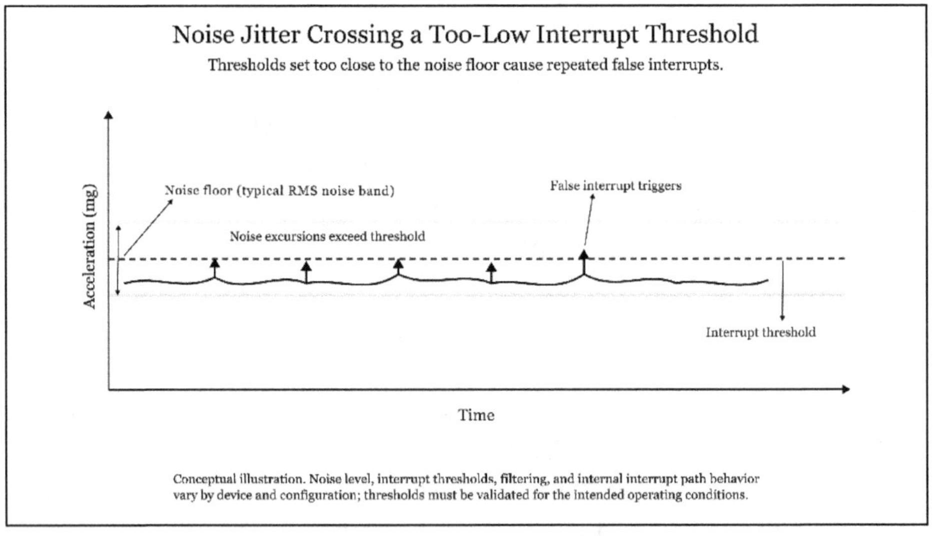

Figure 5.16 Noise jitter crossing a too-low interrupt threshold.

Interrupt behavior is further complicated by internal filtering paths. The signal used to generate interrupts is often filtered differently from the main acceleration output. A datasheet may specify a motion-interrupt low-pass filter fixed at 25 Hz, while the acceleration output bandwidth remains user configurable. As a result, an interrupt may fail to trigger even though the reported output clearly shows motion, or it may trigger even when the reported motion appears weak. Without understanding these separate internal paths, engineers may incorrectly conclude that the sensor is behaving inconsistently, when the real issue is that the interrupt logic and the output stream are not using the same signal path.

Another common assumption is that interrupt timing matches the configured ODR. This is often not true. An accelerometer configured with an ODR of 200 Hz may evaluate interrupt conditions internally at only 50 Hz. Fast transient events can be missed, and interrupt latency may be longer than expected. This behavior is rarely emphasized in datasheets but can significantly affect time critical applications.

Using the same threshold on all axes is another frequent mistake. Axes aligned close to gravity naturally exhibit higher baseline variation than horizontal axes. Applying a single threshold to all axes may produce false positives on one axis and missed events on another. Using vector magnitude or axis specific thresholds improves consistency when device orientation varies.

Tap and double-tap detection are especially sensitive to configuration and mechanical conditions. These features depend on mounting stiffness, enclosure materials, orientation, bandwidth, and threshold tuning. Relying on default register values is rarely sufficient. A tap threshold that works on a rigid evaluation board may fail completely once the sensor is placed in a flexible enclosure or worn on the body. Tap detection almost always requires empirical tuning under real conditions.

Threshold behavior is also affected by hysteresis and debounce logic. Hysteresis is implemented by configuring the interrupt logic to use two thresholds rather than one: one level to trigger the event and another level to clear it. In some sensors, this behavior is built into the internal interrupt engine, while in other systems it is implemented in the host microcontroller software. This separation prevents small fluctuations near the threshold from causing the signal to cross back and forth repeatedly and generate multiple interrupts even though no clear new event has occurred. For example, a threshold set at 40 mg with no hysteresis may produce repeated interrupts during vibration or weak borderline motion. Debounce logic improves reliability further by requiring a minimum time gap between one interrupt and the next.

Another common pitfall is misinterpreting threshold units. Interrupt thresholds are often specified in raw register units rather than physical units. In that case, the number written into the threshold register is not the threshold in mg directly. Instead, each register count corresponds to a fixed acceleration step. For example, if one register LSB corresponds to 16 mg at a given full-scale range, then a threshold register value of 5 means 5 counts of 16 mg each. The actual threshold is therefore $5 \times 16\,mg = 80\,mg$, not $5\,mg$. Failing to convert register units into physical units correctly can produce thresholds that are off by large factors and lead to confusing results.

Interrupt performance must be validated under real operating conditions. Bench testing rarely reflects real-world motion. A fall detection threshold tuned during desk testing may fail when worn by an elderly user because body dynamics, orientation, and timing differ significantly. Motion patterns, noise, and interrupt timing all change in real use.

Misusing thresholds and interrupts usually appears as unreliable event detection rather than outright failure. Systems may trigger false events, miss real ones, drain more power than expected, or behave differently across users and environments. In many cases, the real issue is not the sensor itself, but an oversimplified understanding of how the interrupt engine actually works.

5.10 Mistaking Noise Spikes for Motion

A common mistake, especially in early designs, is treating isolated spikes or single samples as genuine motion events. Accelerometers operate at high sensitivity and sampling rates, so occasional outliers caused by noise, quantization, power disturbances, communication glitches, or environmental vibration are unavoidable. Interpreting these artifacts as real motion leads to false step counts, incorrect fall detection, unstable gesture recognition, and unpredictable system behavior. Reliable motion detection depends on recognizing patterns over time, not reacting to individual samples.

Noise spikes are usually distinguished by their short duration, isolation, and lack of consistent temporal structure. Unlike true motion events, they do not repeat in a physically meaningful sequence and often appear as abrupt excursions that are not supported by neighboring samples. This makes them especially dangerous in systems that rely on simple thresholds or single-sample decisions.

A frequent error is defining events from a single threshold crossing. If one acceleration sample briefly exceeds a threshold, the system may register a step or impact even though no real motion event has occurred. Single-sample decisions are inherently fragile and often produce false detections.

Data integrity issues can produce similar symptoms. Occasional I^2C or SPI read errors may return impossible values, such as measurements that exceed the configured full-scale range. For example, a sensor configured for $\pm 4\ g$ may briefly report values above this limit, such as $+5\ g$ or $+6\ g$. This is not physical motion, but a data error. Treating such values as real events introduces false detections and erratic behavior. Rejecting values outside the valid physical range and validating multi-byte reads helps prevent this class of error.

Short impulses can also be misleading. Brief contacts occur when a device is set down, tapped accidentally, or exposed to nearby vibration. These impulses generate sharp spikes but lack the temporal structure of meaningful motion. A quick bump on a table produces a narrow acceleration spike, but it does not resemble a step, fall, or gesture sequence. Treating it as such leads to false positives.

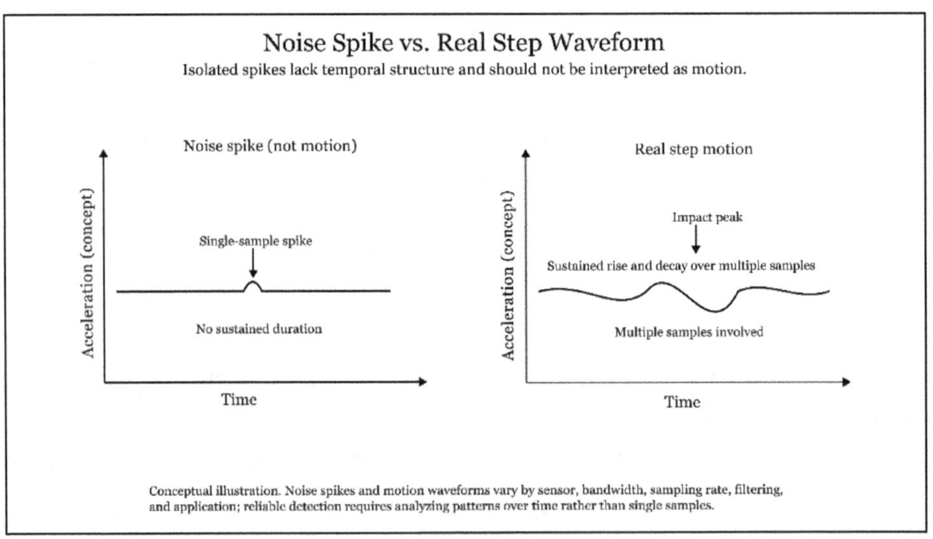

Figure 5.17 Noise spike versus real step waveform comparison.

Figure 5.17 contrasts an isolated noise spike with a genuine step-related acceleration waveform. Noise spikes are brief, single-sample or short-duration excursions that lack consistent shape, duration, or temporal structure. By comparison, a real step produces a structured waveform spanning multiple samples, with a characteristic rise, impact peak, and recovery phase. Treating isolated spikes as motion events leads to false detections, while robust motion detection relies on recognizing patterns that persist over time rather than reacting to instantaneous amplitude alone.

Robust systems reduce false detections by checking whether the signal persists and follows an expected pattern over time rather than relying on instantaneous amplitude alone. Requiring multiple consecutive samples above a threshold and verifying a plausible rise-and-fall behavior improves reliability. Using vector magnitude instead of individual axes can further help when noise affects axes unevenly, although magnitude alone is not sufficient without temporal pattern checks.

This difference is clear when comparing simple and robust algorithms. A simplistic step detector increments a counter whenever acceleration magnitude exceeds a threshold, which fails easily in noisy environments. A more robust method verifies that the signal persists for a short window, returns toward baseline within an expected interval, and matches realistic timing between steps. Noise spikes fail these checks because they lack duration and repeatability.

The same logic applies to fall detection. Systems that rely only on peak acceleration can generate false alarms. A true fall usually includes a short period of near-zero acceleration, followed by a high-impact peak and subsequent stillness. Noise spikes do not follow this sequence and can often be rejected by temporal logic.

Mistaking noise spikes for motion usually shows up as false detections, missed events, or unstable behavior rather than immediate failure. In many cases, the underlying problem is overly simplistic event-detection logic that reacts to isolated spikes instead of validating temporal structure.

5.11 Mistaking Offset Drift for Failure

Offset drift is frequently misinterpreted as sensor failure when observing accelerometer data in real systems. A common mistake is treating a slow change in the zero-g output over time or temperature as evidence of damage or malfunction. In most cases, this behavior is normal and expected. Offset drift results from temperature variation, mechanical stress, and internal stress relaxation, not from loss of sensitivity or degradation of the sensing structure. The physical origins and specification-level treatment of offset drift are discussed in earlier chapters and should be referenced for a detailed explanation.

Engineers often encounter this issue when a device that was previously stable at rest begins reporting a slightly different baseline value after warm-up, environmental temperature change, enclosure assembly, or extended operation. Interpreting this shift as a hardware fault leads to unnecessary debugging, component replacement, or attempts to fix the behavior using excessive filtering or algorithm changes.

Another frequent error is expecting different devices to exhibit identical offsets. Device-to-device variation is normal, even within datasheet limits. A stationary device will not necessarily read zero on all axes, because one or more axes may include a gravity component depending on orientation.

True hardware failure presents differently. Abrupt offset changes of large magnitude without an external cause, rapidly increasing noise, failed self-test responses, or unstable communication are stronger indicators of actual damage or malfunction.

Offset drift should be treated as a predictable characteristic and handled through calibration or compensation rather than interpreted as failure. When misdiagnosed, offset drift is often blamed on the sensor, when the real issue is an incorrect expectation of what normal behavior looks like.

5.12 Treating Accelerometers in Isolation

A common system-level mistake is assuming that an accelerometer alone can fully describe motion, orientation, or device behavior. Accelerometers measure linear acceleration, including gravity, very well, but they do not directly measure rotation rate, absolute heading, or position. Treating an accelerometer as a complete motion sensing solution leads to unstable estimates, drift, and incorrect interpretation of real-world events. Reliable systems treat accelerometers as one component within a broader sensing and interpretation framework.

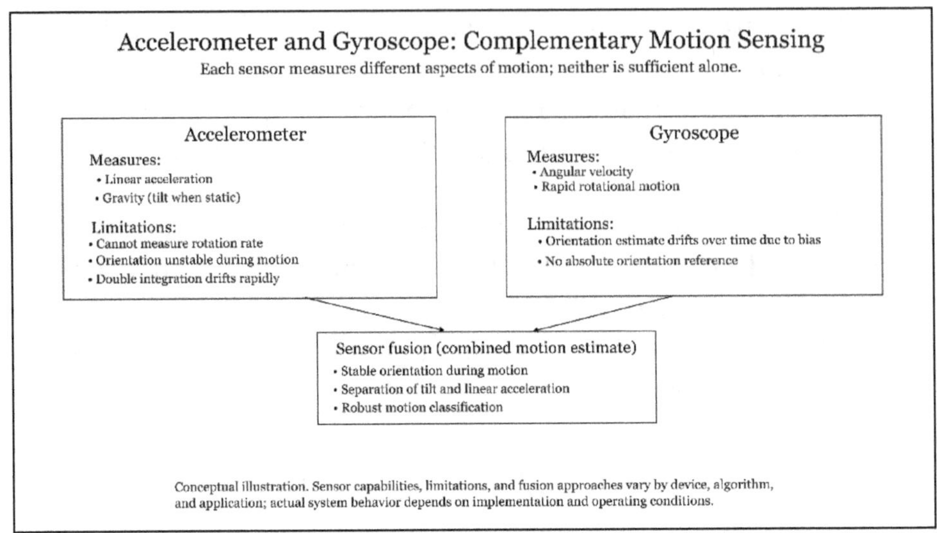

Figure 5.18 Accelerometer and gyroscope complementing each other.

Figure 5.18 illustrates how accelerometers and gyroscopes provide complementary information about motion. Accelerometers measure linear acceleration and gravity, making them useful for detecting tilt and low frequency motion when the device is relatively static.

One frequent misconception is expecting accelerometers to measure rotation. Accelerometers do not measure angular velocity. Rotation only appears indirectly when it changes the direction of gravity relative to the sensor or produces centripetal acceleration. Rotating a device rapidly about its center may produce little change in measured acceleration, even though angular velocity is high. A gyroscope clearly detects this motion, while the accelerometer may not. Using accelerometer data alone to infer rotation rate leads to missing or ambiguous information.

Gyroscopes measure angular velocity and capture rapid rotational motion but suffer from drift when integrated over time. When used together through sensor fusion, these sensors compensate for each other's weaknesses, enabling stable orientation estimates, separation of tilt from dynamic acceleration, and robust motion interpretation. Treating an accelerometer in isolation ignores these complementary roles and leads to unstable or incomplete system behavior.

Another common mistake is expecting stable orientation estimates during dynamic motion. Accelerometers can estimate tilt relative to gravity only when dynamic acceleration is small. During walking, running, vibration, or rapid movement, linear acceleration overwhelms gravity, making tilt estimates unreliable. Because an accelerometer cannot distinguish gravity from other linear acceleration or measure angular rate, accurate dynamic orientation tracking requires a gyroscope.

Heading estimation is another area where accelerometers are often misused. Accelerometers cannot determine compass heading. Absolute heading requires sensing the Earth's magnetic field. A robot or handheld device using only an accelerometer and gyroscope may estimate orientation initially, but heading drifts over time due to gyroscope integration error. Without a magnetometer or external reference, absolute direction is lost.

Attempting navigation using accelerometer data alone leads to even more severe failure. Position is obtained by double integration of acceleration:

$$position = \int \int acceleration \, dt^2$$

Even very small offset or noise errors grow rapidly during integration, causing position estimates to diverge within seconds. Accelerometers alone are therefore unsuitable for navigation without additional references such as gyroscopes, magnetometers, or external position sensors.

Motion interpretation also depends heavily on context. Signals vary based on how and where the sensor is mounted and how the system is used. A vibration level that is normal for one machine may indicate a fault in another. A wrist mounted wearable produces very different motion patterns than a waist mounted device, even for the same activity. Without context, raw acceleration data lacks meaning.

Treating accelerometers in isolation often leads to systems that drift, behave inconsistently, or work only under controlled test conditions. The root problem is usually not a hardware limitation, but an incomplete sensing strategy that ignores how motion, orientation, filtering, timing, and system context interact.

5.13 Startup and Initialization Errors

A frequently overlooked mistake is assuming that accelerometer data is valid immediately after powerup or configuration changes. In reality, accelerometers require time to stabilize during startup and after updates to power modes, bandwidth, ODR, or filtering. Internal biasing, reference voltages, digital filters, and clocking must all settle before measurements become reliable.

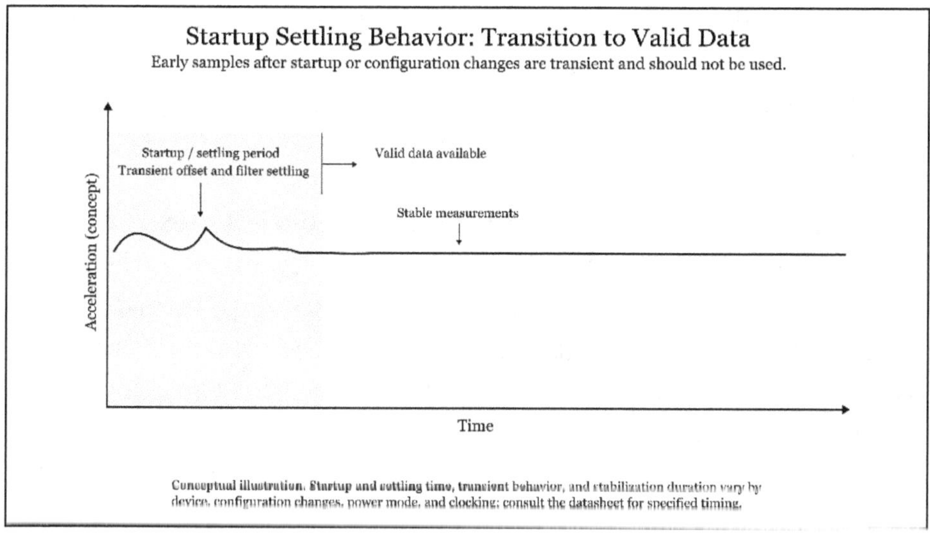

Figure 5.19 Startup settling behavior and transition to valid accelerometer data.

Figure 5.19 illustrates the transient behavior that occurs immediately after accelerometer startup or configuration changes. During the initial settling period, internal biasing, reference voltages, and digital filters stabilize, causing output values to fluctuate or ramp toward their correct level.

These early samples do not represent valid motion and should be discarded. Once settling is complete, the output stabilizes and can be used reliably for calibration, motion detection, and algorithm initialization. Ignoring this transition often leads to false events, incorrect baselines, or unstable system behavior. Immediately after enabling active mode, an axis may report a significant offset that disappears after a few milliseconds. Using these early samples for motion detection or calibration can trigger false events or corrupt baseline estimates.

Startup behavior is predictable once understood, but when ignored it is easily misinterpreted as noise, drift, or even hardware failure. Datasheets specify a startup or turn-on time between enabling active mode and when valid data becomes available. During this interval, early samples may contain large offsets or transient behavior.

Configuration changes introduce similar issues. Changing bandwidth, ODR, or filter settings resets internal filter states. During the settling period, output data may ramp toward the correct value rather than settling instantly as the internal filters converge. Treating these transient samples as valid can produce incorrect offset estimates and unstable algorithms. Allowing several samples to pass before trusting the data avoids this problem.

Another common error is assuming that configuration persists across power cycles or deep sleep. Many accelerometers revert to default settings after power loss. After wake-up, a device may return to its default full-scale range, ODR, or filter configuration. If software continues using scaling factors from the previous configuration, such as the counts-to-g conversion associated with the earlier full-scale range, sensitivity setting, or output format, all measurements are incorrect. Robust systems explicitly reconfigure all required registers after every startup or wake-up.

Thermal stabilization also matters. Internal circuitry generates heat after power-on, and for precision applications the accelerometer output may drift until thermal equilibrium is reached. An offset measured immediately after startup may differ from the value observed after seconds or minutes of operation, depending on the device and system conditions. Calibrating too early captures transient behavior rather than steady-state conditions.

Proper data handling matters as well. If samples are read without checking data ready flags or status registers, old or partially updated values may be processed. Reading faster than the

ODR returns repeated samples, creating the impression of frozen or delayed motion. Proper synchronization with data ready signals prevents this class of error.

Startup and initialization problems usually appear during the first moments of operation, when systems may trigger false motion events, show unstable baselines, or behave inconsistently. In many cases, the real issue is simply that the data is being read or used before the sensor has fully initialized and settled.

5.14 Misusing Self-Test

A common mistake when working with accelerometers is misusing or misinterpreting the built-in self-test feature. Self-test is frequently treated as a calibration tool or as a guarantee of overall sensor accuracy. This is incorrect. Self-test is a diagnostic check that verifies basic mechanical and electrical integrity, not offset accuracy, sensitivity calibration, axis alignment, noise performance, or long-term stability. The operating principle of self-test and the expected response behavior are illustrated in Chapter 3 (see **Figure 3.12** and **Figure 3.13**.

One common misuse is interpreting near-limit self-test responses as failures. Datasheet self-test ranges intentionally include tolerance bands to account for process variation, temperature, configuration, and operating conditions. As shown in **Figure 3.13**, responses near the minimum or maximum specified limits are still valid when they fall within the datasheet-defined range.

Another frequent error is ignoring configuration dependence. Self-test response varies with full-scale range, bandwidth, ODR, power mode, and axis. Comparing results obtained under one configuration against datasheet limits defined for another leads to false failure conclusions.

Running self-test under motion, vibration, or mechanical stress is also a common mistake. External acceleration contaminates the measurement and can push results outside the expected range even when the sensor is healthy. Self-test should be performed only when the device is stationary and mechanically stable.

Self-test is sometimes enabled during normal operation without masking or pausing downstream processing. Because it intentionally alters the accelerometer output, failing to isolate this interval can trigger false motion events, unstable filters, or incorrect algorithm behavior.

During initialization, self-test introduces a different risk. It temporarily shifts the accelerometer output, so data captured during or immediately after self-test does not represent normal sensor operation. Calibration and algorithm initialization should therefore begin only after self-test has completed and the output has returned to normal.

Passing self-test does not imply correct offset, sensitivity, or temperature compensation. A device can pass self-test while still producing inaccurate results if calibration is incomplete or missing.

Self-test should be treated strictly as a basic integrity check and interpreted only within datasheet limits and test conditions. Misinterpreting the result often leads to incorrect conclusions, when the real issue is usually improper test execution or expectations that do not match the datasheet definition of self-test.

5.15 Overreliance on Raw Data

A common mistake in accelerometer-based system design is assuming that raw acceleration values can be interpreted directly without additional processing. Raw data contains noise, gravity, orientation effects, and transient artifacts that must be filtered, transformed, or combined with other information before meaningful conclusions can be drawn.

Raw data is a starting point, not a finished signal. Even when a device is motionless, raw accelerometer outputs fluctuate. At rest, a sensor may report $a_x = +0.008\ g$, $a_y = -0.015\ g$, and $a_z = +1.012\ g$. These variations are normal and reflect noise and finite digital resolution, not sensor malfunction. Without filtering, such fluctuations lead to jittery tilt estimates, unstable gesture detection, and erratic thresholds.

Another frequent mistake is ignoring the need to separate gravity from dynamic motion. Raw acceleration always includes gravity, and changes in orientation redistribute this component across axes. Tilting a stationary device causes a_z to decrease while a_x or a_y increases. This is not dynamic movement, but a change in orientation relative to gravity. During walking, forward acceleration further alters the measured acceleration vector. Without separating gravity, algorithms confuse orientation changes with actual motion.

Raw accelerometer data becomes even harder to interpret during dynamic motion because gravity and linear acceleration are combined in the same measurement. During walking, running, or vibration, motion-induced changes can be mistaken for changes in tilt if the signal is interpreted without filtering, compensation, or context. In these situations, accelerometer data must be processed carefully and often combined with other information before reliable conclusions can be drawn.

Analyzing only a single axis introduces another layer of error. Many real-world motions involve all three axes simultaneously. A gesture may include vertical motion, lateral sweep, and rotation at the same time. Observing only one axis hides important information and weakens classification accuracy. Using vector magnitude or cross-axis features improves robustness across orientations.

Context is equally important. Acceleration values have meaning only within the system and processing chain that interprets them. Raw acceleration samples do not directly reveal motion intent, activity type, or mechanical state without filtering, feature extraction, or application-specific interpretation. Interpreting raw values without understanding mounting location, usage patterns, and physical constraints leads to incorrect conclusions.

In addition, raw data must be calibrated before interpretation. If the Z-axis reports 1.07 g at rest due to offset or sensitivity error, tilt estimation is off by several percent. Without calibration and compensation, even simple applications become unreliable. Here, calibration means measuring and correcting sensor-specific errors such as offset and sensitivity, while compensation means correcting the output further when operating conditions, especially temperature, change.

Overreliance on raw data usually appears as noisy, inconsistent, or fragile system behavior, especially under real-world conditions. In most cases, the missing piece is not better hardware, but more appropriate processing, sensor fusion, or context-aware interpretation.

5.16 Timing and Synchronization Errors

Timing issues are among the most frustrating problems in systems that use accelerometers because they often appear random. Symptoms include delayed response, unstable control loops, inconsistent gesture recognition, and vibration spectra that never quite match expectations. These problems usually stem from misunderstanding latency, sampling timing, filter delay, FIFO buffering, or synchronization between sensors and processing logic.

Accelerometers do not deliver instantaneous measurements. Internal filtering, buffering, and digital processing all introduce delays. If these delays are not understood and managed explicitly, system behavior becomes difficult to predict.

As discussed earlier, a common error is reading data without verifying freshness. Accelerometers update output at the configured ODR. Reading faster than that rate often returns the same sample multiple times. If the ODR is 100 Hz and the microcontroller reads data every 2 ms, several reads return identical samples. Treating these repeated values as new motion introduces apparent lag and distorts timing. Using data-ready flags, interrupts, or FIFO status indicators ensures that each processed sample is a new measurement.

Timing problems also arise when multiple sensor data streams are used together. If those streams are not properly time-aligned, fusion algorithms behave poorly. Combining accelerometer data with other sensor inputs without proper resampling or timestamp alignment can produce oscillation, drift, or unstable estimates. Consistent sampling or accurate timestamps are therefore essential for reliable sensor fusion.

Figure 5.20 illustrates a common timing error where the microcontroller reads accelerometer data faster than the sensor updates its output. Accelerometers update measurements at the configured ODR, not on every read. When the MCU reads more frequently than this rate, the same sample is returned multiple times. Treating these repeated values as new motion introduces apparent lag, distorts timing, and corrupts downstream processing. Verifying data freshness using data-ready flags, interrupts, or FIFO status ensures that each processed sample represents a new measurement.

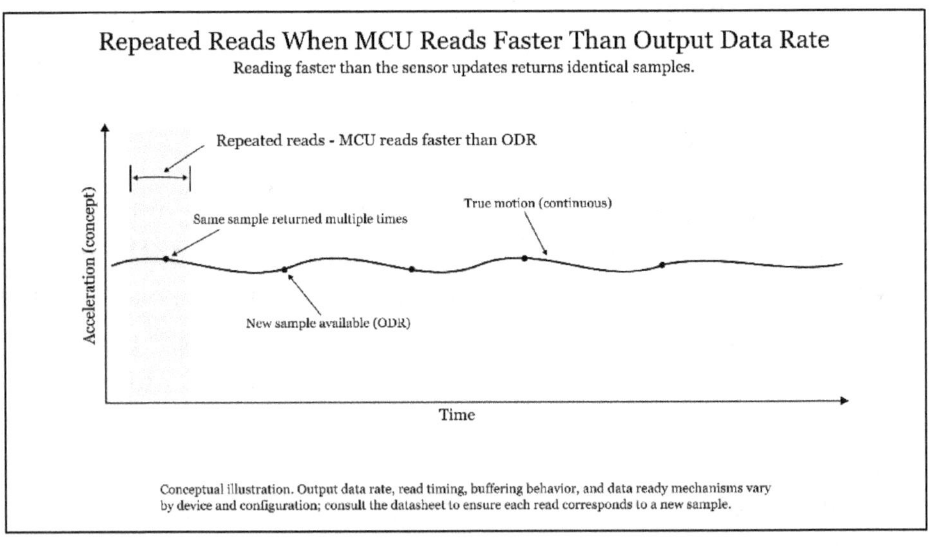

Figure 5.20 Repeated reads returning identical samples when the MCU reads faster than the ODR.

Filter latency is another frequent source of confusion. Both analog and digital filters introduce delay between physical motion and reported data, and this delay increases as cutoff frequency is reduced. This occurs because a lower cutoff frequency forces the filter to respond more gradually, suppressing fast changes more strongly and causing the output to take longer to follow the input. A low-pass filter with a cutoff near 10 Hz may introduce tens of milliseconds of delay. In wearables this may be acceptable, but in drones or robotics it can destabilize control loops if not accounted for. Filter delay must be included when tuning timing-sensitive algorithms.

FIFO buffering introduces additional latency. FIFOs reduce bus traffic and save power, but samples stored in the buffer are, by definition, delayed. If a FIFO holds 32 samples and the ODR is 100 Hz, the oldest sample is delayed by about 320 ms. For real-time control or fast gesture detection, this delay is unacceptable. FIFO depth must be chosen based on acceptable latency, not convenience.

Interrupt latency is another commonly overlooked factor. Interrupts are not instantaneous, as internal filtering, synchronization logic, and microcontroller handling all introduce delay. An interrupt configured for a 100 Hz event may arrive several milliseconds after the physical motion occurs. Assuming zero latency leads to timing errors in gesture detection and control logic, so interrupt timing must be measured and accounted for in system design.

Communication bus latency adds further delay. I^2C and SPI transfers consume time, especially when reading multiple registers. At 400 kHz I^2C, reading multiple bytes may take 1 millisecond or more. At high ODRs, this consumes a significant fraction of the sampling period and increases effective latency. High-rate systems benefit from faster buses or controlled FIFO usage.

Some accelerometers sample internally at a higher rate and then decimate to the configured ODR. This internal processing introduces delays that is not always obvious from register settings alone. Without understanding this behavior, engineers often mistune control and detection algorithms.

Timing and synchronization errors rarely show up in static testing. They dominate dynamic behavior. Systems that feel sluggish, unstable, or out of sync often suffer not from noise or calibration problems, but from unaccounted latency and misaligned timing.

5.17 Takeaways

In this chapter, we examined the most common mistakes engineers make when working with accelerometers. These mistakes do not occur because accelerometers are unreliable, but because they are physical devices whose behavior depends strongly on configuration choices, mechanical conditions, timing, and how motion data is interpreted over time. Understanding these pitfalls is essential for building systems that behave predictably outside controlled test environments.

We began with foundational configuration errors, including incorrect full-scale range selection, misunderstanding noise behavior, confusing bandwidth with ODR, and inadequate filtering choices. These mistakes degrade signal quality by reducing effective resolution, increasing noise, introducing aliasing, or distorting motion before algorithms ever see the data.

We then explored errors related to power modes, orientation, axis alignment, and calibration. Misusing low-power modes, misinterpreting axis orientation, skipping or misapplying calibration, and misunderstanding temperature effects introduce subtle but persistent errors in offset, sensitivity, and orientation that are often mistaken for sensor defects.

Mechanical and environmental factors emerged as another major source of failure. Improper sensor placement, enclosure stress, PCB flexing, vibration coupling, and thermal effects can all shift sensor behavior over time. These issues frequently appear as software bugs or algorithm instability, leading engineers to debug the wrong part of the system.

We also addressed timing and system-integration errors. Startup behavior, filter settling, FIFO buffering, interrupt latency, communication delays, and synchronization mismatches introduce hidden delays that undermine real-time assumptions. When these effects are ignored, systems become sluggish, unstable, or difficult to reason about.

Finally, we focused on interpretation mistakes in motion data. Treating noise spikes as real motion, relying on single samples instead of temporal patterns, misusing thresholds and interrupts, and over-relying on raw data all lead to systems that fail under real-world variability. Real motion must be interpreted as structured, multi-axis behavior unfolding over time, not as isolated values crossing fixed thresholds.

Taken together, these mistakes point to a central lesson: accelerometers are powerful sensors, but they demand thoughtful configuration, mechanical awareness, disciplined signal processing, and system-level reasoning. When these factors are handled carefully, accelerometers deliver robust and reliable insight. When they are ignored, even high-quality sensors produce confusing and inconsistent results.

With this foundation in place, Chapter 6 - Accelerometer Applications shows how accelerometers are applied successfully across real systems, demonstrating how correct configuration, calibration, filtering, timing, and interpretation come together in wearables, consumer devices, industrial monitoring, robotics, and other practical deployments.

CHAPTER 6
Accelerometer Applications

Accelerometers are used in an astonishing variety of products, from phones and wearables to vehicles, medical devices, industrial machinery, and aerospace systems. Their ability to sense acceleration, vibration, tilt, and gravity makes them one of the most versatile sensors available today. By understanding how accelerometers work and how their specifications relate to performance, engineers can integrate them into systems that range from simple motion detection to sophisticated monitoring and control applications.

With the core concepts, specifications, and common pitfalls established in the previous chapters, we now turn to how accelerometers are used in real systems. Rather than treating accelerometers as generic motion sensors, this chapter examines applications through the lens of system requirements and tradeoffs. The same accelerometer technology can serve very different roles depending on how it is configured, where it is placed, and how its data is processed.

Figure 6.1 groups these domains together and highlights that accelerometers should not be viewed as generic motion sensors, but as configurable system components whose role is defined by the surrounding system and application context.

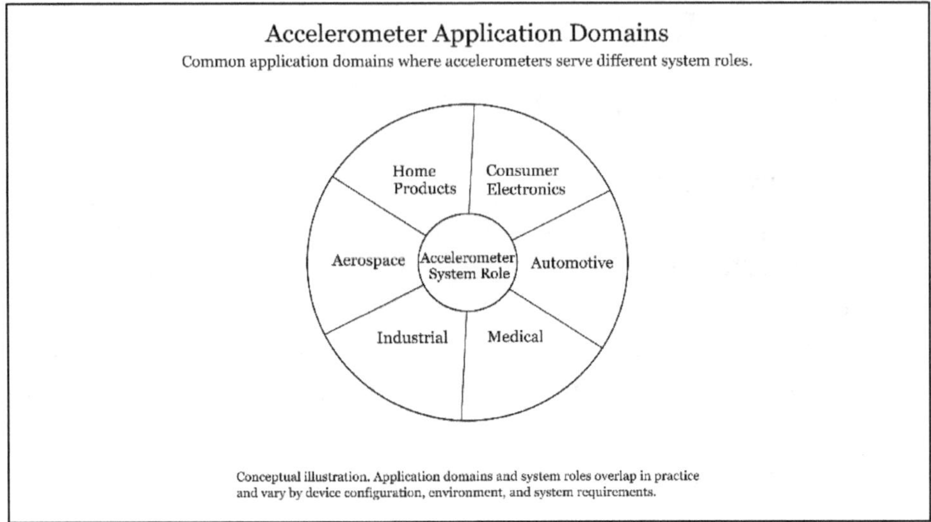

Figure 6.1 Accelerometer application domains across real-world systems.

Accelerometers are used across a wide range of application domains, each imposing distinct system requirements and constraints. While the underlying sensing principle remains the same, differences in environment, reliability expectations, power budgets, and interpretation logic lead to very different design choices.

In practical designs, accelerometers are selected not only for what they measure, but for how reliably and efficiently they deliver that information within a specific environment. Factors such as operating temperature, mechanical stress, vibration exposure, power budget, latency, and long-term stability all influence sensor choice and overall system architecture. As a result, an accelerometer that performs well in one application may be poorly suited for another, even when the underlying sensing principle is the same.

6.1 Consumer Devices

Consumer electronics represent one of the earliest and most widespread markets for MEMS accelerometers. Their small size, low power consumption, and ability to measure both dynamic motion and static gravity made them foundational to smartphones, wearables, and other portable devices. Today, accelerometers play an important role in user interfaces, health monitoring, photography, gaming, and many everyday consumer products, shaping how users interact with modern technology.

Consumer applications emphasize low power operation, compact packaging, and predictable performance under everyday conditions. Devices must operate reliably across a wide range of user interactions, from gentle indoor handling to motion, vibration, and temperature variation encountered during daily use. Although consumer products do not face the extreme environments of industrial or automotive systems, users expect seamless and invisible behavior at all times.

In many consumer products, accelerometers operate quietly in the background to support responsive user interfaces, device automation, basic safety features, and energy-efficient operation. These applications usually do not require extreme precision or wide bandwidth, but they do demand low-cost, robust solutions with reliable threshold detection, long-term stability, and tolerance to routine vibration, occasional shock, and varied mounting conditions.

6.1.1 Smartphones and Tablets

In smartphones and tablets, accelerometers function as motion interpreters that enable intuitive, motion-aware features. Orientation detection relies on measuring the gravity vector so the device can determine whether it is held upright, sideways, or inverted, and rotate the display accordingly. This application depends on low noise, stable zero-g offset, sufficient resolution for fine tilt detection, and sensitivity stability across temperature.

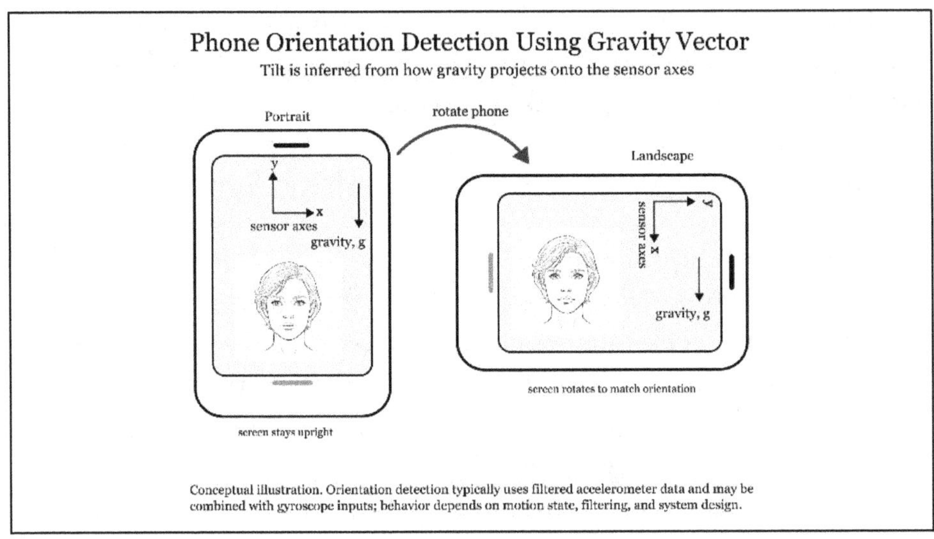

Figure 6.2 Phone orientation detection using gravity projection.

As shown in **Figure 6.2**, smartphone orientation detection is based on how the gravity vector projects onto the accelerometer's sensor axes. When the device is held in portrait or landscape orientation, the direction of gravity relative to the sensor changes, producing different static acceleration components along the X and Y axes. By monitoring these components while the device is stationary or moving slowly, the system can infer tilt and determine the correct screen orientation. Reliable operation depends on low noise, stable zero-g offset, and consistent sensitivity so that orientation transitions occur smoothly without jitter or unintended rotation.

Many devices also implement lift-to-wake or wake-on-motion features. These functions detect characteristic motion patterns when the device is picked up and require ultra-low-power operating modes, programmable thresholds, and fast interrupt response to preserve battery life while remaining responsive.

Step counting and general activity detection extract low frequency motion components associated with walking and daily movement. Reliable performance depends on low noise density, appropriate ODR settings, efficient filtering, and stable sensitivity. Motion-based gestures such as shaking, tapping, or tilting rely on sufficient bandwidth, low latency, and interrupt-based event detection.

6.1.2 Wearables and Fitness Trackers

Wearable devices rely heavily on accelerometers because they must interpret human motion while consuming very little power. Step counting and walking dynamics are derived from repetitive acceleration patterns, typically in the low-frequency range. These applications emphasize low RMS noise, adequate ODR, and long-term sensitivity stability.

Figure 6.3 highlights why fall detection in wearables is challenging from a system-design perspective. The sensor must capture both the brief near-zero-acceleration interval associated with loss of support and the high-impact portion of the event without saturating or missing important detail. Reliable performance also depends on low noise, fast response, appropriate full-scale range, and robustness to device placement, user motion, and false triggers from everyday activity.

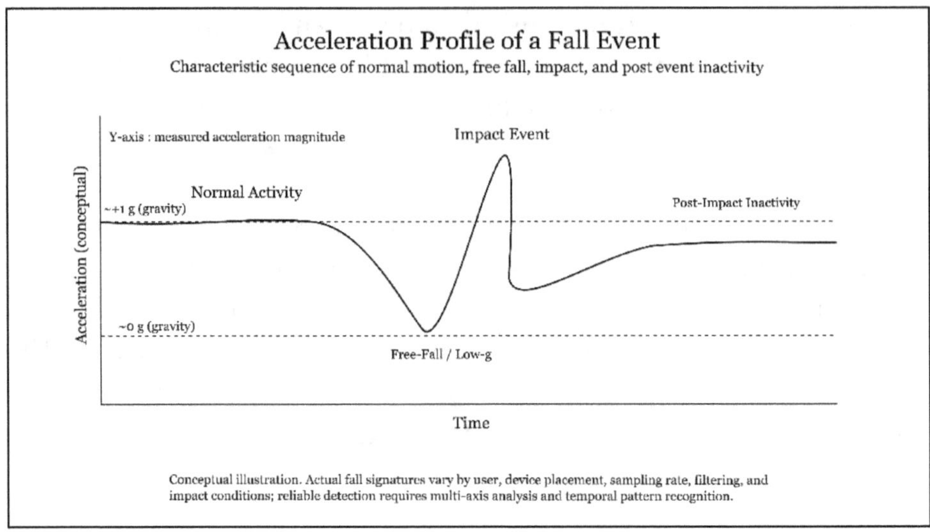

Figure 6.3 Conceptual acceleration profile of a fall event measured by an accelerometer.

Sleep monitoring uses accelerometers to detect restlessness and micro-movements over extended periods. This places strong demands on noise performance, resolution, and offset stability at low frequencies. Fall detection combines sudden acceleration events with periods of inactivity, requiring wider full-scale ranges, accurate threshold detection, and strong shock resistance.

Posture and activity classification distinguish between sitting, standing, walking, and lying down by analyzing gravity orientation and low frequency motion. These functions depend on low noise, stable offset, and accurate tilt measurement across temperature.

6.1.3 Laptops and Portable Computers

Earlier laptop designs used accelerometers for hard disk drive protection. By detecting free fall, the system could move the drive heads to a safe position within milliseconds to prevent damage. Although this application has largely disappeared with solid state storage, it illustrates the need for fast response, reliable freefall detection, and robust shock tolerance.

In convertible laptops and tablets, accelerometers remain useful for adjusting user interface behavior based on orientation, such as switching keyboard or touchpad modes. These features rely on accurate tilt detection and low offset drift to avoid unintended transitions.

6.1.4 Cameras and Handheld Video Devices

Accelerometers assist cameras and handheld video devices by detecting translational motion and vibration. When combined with gyroscopes, they help stabilize images and video during handheld operation. These applications emphasize low noise, controlled bandwidth, and minimal temperature drift.

Some cameras also use accelerometers to provide level indicators that help users align shots horizontally. This function relies on accurate static gravity measurement and stable offset behavior.

6.1.5 Gaming, Virtual Reality (VR), and Augmented Reality (AR)

Gaming controllers and VR or AR headsets use accelerometers to track motion, gestures, and orientation. Swinging, shaking, or tilting a controller produces characteristic acceleration patterns that are interpreted as user input. These applications require moderate bandwidth, low latency, and consistent ODR.

As shown in **Figure 6.4**, accelerometers in VR controllers measure linear acceleration along multiple directions as the controller is lifted, swung, or moved laterally. These acceleration vectors vary over time and form characteristic patterns that are interpreted as gestures rather than as absolute position. Orientation changes are inferred from gravity projection and refined using gyroscope data, while accelerometer measurements provide responsiveness to dynamic motion. This combination enables low latency, intuitive interaction in gaming and immersive systems while maintaining low power consumption.

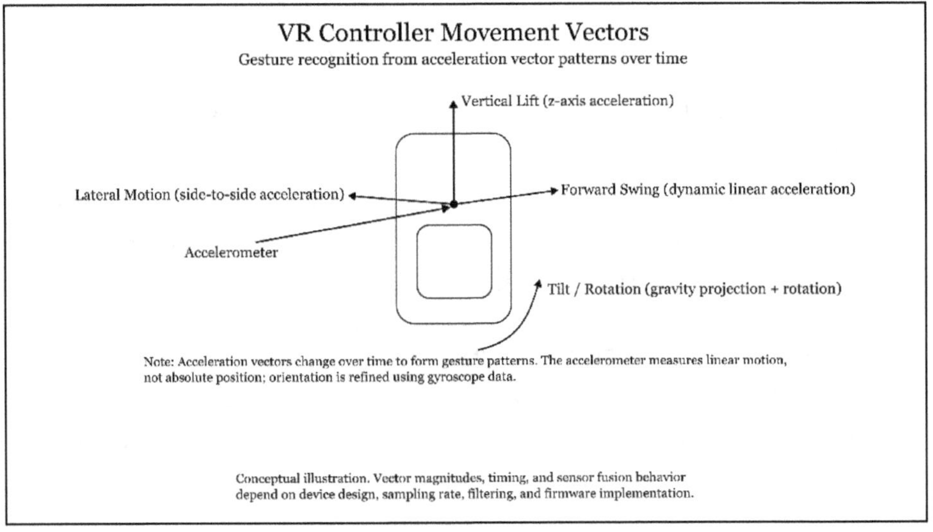

Figure 6.4 Gesture interpretation using acceleration vectors in a VR controller.

In six-degree-of-freedom tracking systems, accelerometers operate alongside gyroscopes and magnetometers to support motion tracking and spatial awareness. Synchronized sampling, low noise, and stable offsets are important for maintaining accuracy and reducing drift over time.

6.1.6 Smart Home Devices

Many smart home products integrate accelerometers to improve usability and automate behavior. Robotic vacuum cleaners detect collisions, slopes, and stuck conditions using motion cues. Some smart thermostats and smart displays can use movement-based signals (including accelerometers in certain designs) to infer interaction or occupancy.

Smart-home designs typically use accelerometers in simple, low-power event-detection roles rather than for detailed motion measurement. The goal is often to detect taps, lifts, knocks, tilt changes, or abnormal movement reliably enough to trigger the correct system response without adding noticeable cost or power burden.

These applications emphasize low power consumption, reliable threshold detection, and robust shock tolerance rather than extreme measurement precision.

6.1.7 Household Appliances

Washing machines commonly use accelerometers to detect unbalanced loads during spin cycles. Excessive vibration indicates uneven distribution of clothing, prompting the control system to reduce speed, redistribute the load, or pause operation. These functions rely on bandwidth suitable for vibration detection, moderate noise performance, and strong mechanical durability.

As shown in **Figure 6.5**, the accelerometer is used to translate excess vibration during the spin cycle into a control response. Rather than measuring motion for detailed analysis, the system looks for imbalance-related vibration patterns severe enough to trigger corrective action, such as reducing spin speed, redistributing the load, or pausing operation. This application emphasizes robust vibration detection, reliable threshold behavior, and mechanical durability rather than high-resolution motion measurement.

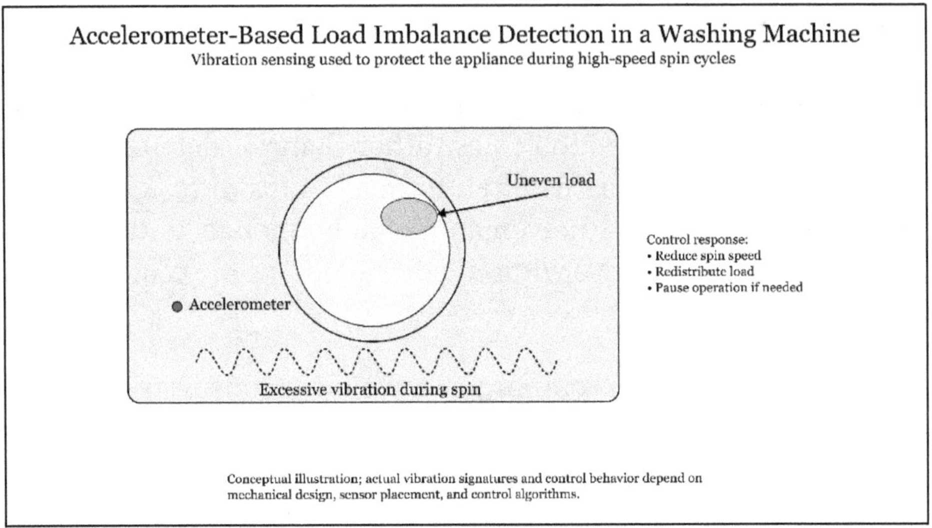

Figure 6.5 Accelerometer-based detection of load imbalance in a washing machine.

Dishwashers, refrigerators, and other household appliances may use accelerometers to detect door motion, monitor compressor vibration for diagnostics, or provide tilt-related safety functions in portable designs. In these cases, the accelerometer serves as a simple embedded monitor that supports reliable operation under everyday mechanical conditions.

6.1.8 Tools and Power Equipment

Power tools and professional equipment integrate accelerometers to improve safety and usability. Some tools detect sudden reactive forces associated with kickback or binding. When abnormal acceleration is detected, the system can shut down the motor to reduce injury risk. These applications require fast response, high shock tolerance, and reliable threshold behavior.

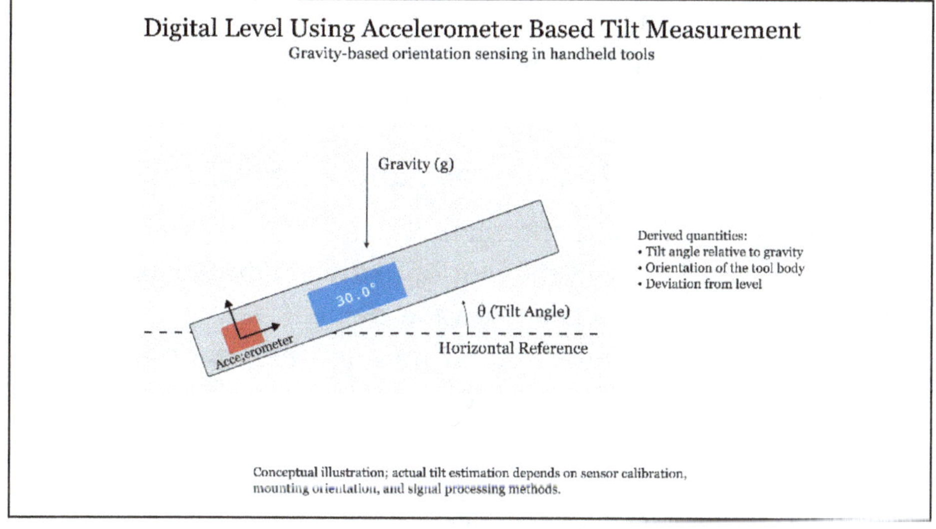

Figure 6.6 Digital leveling tool using accelerometer-based tilt measurement.

As shown in **Figure 6.6**, handheld digital leveling tools use accelerometers to measure tilt by resolving the direction of the gravity vector relative to the tool body. By projecting gravitational acceleration onto the sensor axes, the system computes deviation from level and displays tilt angle in real time. These applications emphasize stable offset behavior, low noise for precise readings, and accurate gravity-based orientation measurement rather than high dynamic range or bandwidth.

Related products such as laser guides and compact layout tools use the same gravity-based tilt principle for alignment and setup. In these designs, stable offset and low noise matter because small angular errors can become visible in the final reading.

6.1.9 Home Safety and Security Devices

Accelerometers enhance safety features in products such as smart smoke detectors, garage door openers, stairlifts, and home security sensors. They can detect abnormal motion, impacts, tilting, or tampering events. Reliable detection with minimal false alarms is essential, particularly in always-on systems where stability over time matters more than raw sensitivity.

6.1.10 Toys, Remote Controls, and Consumer Gadgets

Accelerometers add interactivity to toys, remote controls, and small consumer gadgets. Motion-activated toys, gesture-controlled remotes, and shake-to-wake devices rely on consistent threshold detection and minimal power consumption. In these designs, simplicity and robustness are more important than fine resolution.

6.2 Automotive Systems

Automotive systems impose some of the most demanding requirements on accelerometers. Unlike consumer electronics, vehicles operate across wide temperature ranges, experience continuous vibration, and must remain reliable for many years without recalibration. Accelerometers support safety-critical and performance-related functions, making predictability, robustness, and long-term stability just as important as nominal accuracy.

In automotive applications, accelerometers are evaluated not only by datasheet specifications, but by how consistently those specifications hold under mechanical stress, electrical noise, and temperature extremes. Parameters such as offset drift, noise, bandwidth, shock survivability, and temperature stability directly influence system safety and reliability.

6.2.1 Airbag and Crash Detection Systems

Airbag deployment was one of the earliest large scale automotive uses of accelerometers. During a crash, the sensor must detect a rapid acceleration pulse and trigger the restraint system within milliseconds. This requires extremely fast response, very high full-scale range, low latency signal processing, and carefully tuned filtering to avoid false activation.

As shown in **Figure 6.7**, a vehicle crash produces a short duration acceleration pulse that must be detected and interpreted within milliseconds. Airbag deployment is not triggered by a single acceleration peak alone, but by evaluating whether the crash pulse exceeds a defined threshold for a sufficient duration within a decision window. This requires accelerometers with very high bandwidth, wide full-scale range, and low latency signal paths to capture sharp impact profiles accurately. Robust filtering and predictable sensor behavior are essential to ensure rapid deployment during severe collisions while avoiding false activation during non-crash events.

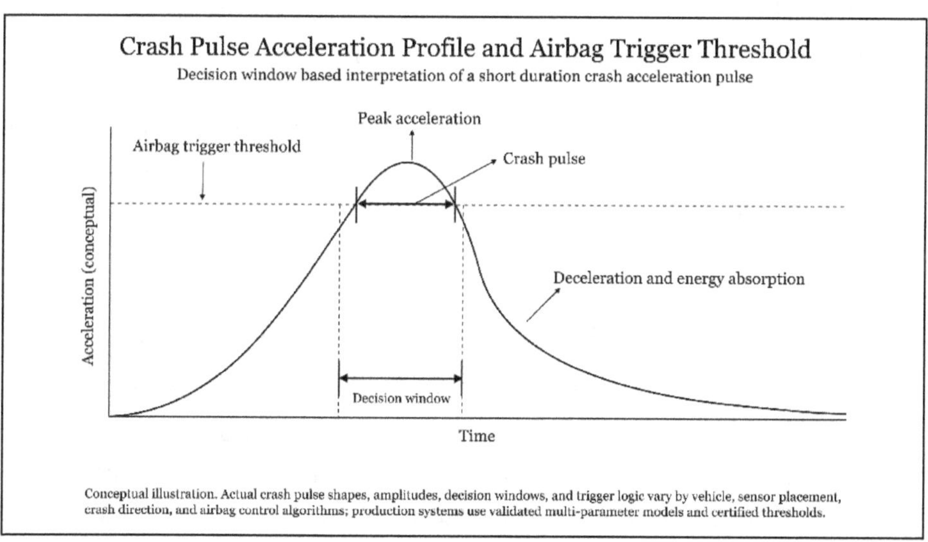

Figure 6.7 Conceptual crash acceleration pulse with airbag trigger threshold.

Accelerometers used as crash sensors must also survive extreme mechanical shock, often thousands of g, while maintaining predictable behavior across the full automotive temperature range. Bandwidth and ODR must be high enough to capture sharp impact profiles without distortion.

6.2.2 Electronic Stability Control and Rollover Detection

Electronic Stability Control systems use accelerometers to measure lateral and longitudinal vehicle motion associated with skidding, loss of traction, or rollover risk. These measurements are combined with gyroscope data, wheel speed sensors, and steering angle information to make real-time control decisions.

Low noise is essential for detecting subtle changes in vehicle dynamics, while low offset drift is critical because bias errors translate directly into incorrect stability assessments. Moderate bandwidth is typically sufficient, but repeatability and temperature stability over the vehicle lifetime are crucial.

Rollover detection relies on monitoring the relationship between gravitational acceleration and lateral forces, making accurate multi-axis measurement and stable sensitivity especially important.

6.2.3 Inertial Navigation and Dead-Reckoning

Accelerometers form part of inertial measurement units used for dead-reckoning navigation when Global Positioning System (GPS) signals are unavailable, such as in tunnels, parking structures, or dense urban environments. In these systems, accelerometer errors accumulate over time, so bias stability and noise density strongly influence navigation accuracy.

Key requirements include stable offset behavior, predictable sensitivity across temperature, synchronized sampling with gyroscopes, and consistent startup performance. Although automotive inertial measurement units (IMUs) do not usually require aerospace-grade accuracy, improvements in MEMS accelerometers have made them increasingly practical for mid-range navigation systems.

6.2.4 Ride Comfort and Suspension Systems

Some vehicles use accelerometers to monitor chassis motion and suspension behavior. These measurements support active suspension, adaptive damping, road condition detection, and body roll estimation.

Bandwidth must be adequate to capture road induced vibrations, while linearity and cross-axis accuracy influence control precision. These systems prioritize repeatability and consistency over extreme sensitivity, since relative measurements are often more important than absolute accuracy.

6.2.5 Electric Vehicle Motor and Drivetrain Monitoring

Electric vehicles increasingly use accelerometers for vibration-based diagnostics of motors and drivetrains. These sensors detect imbalance, bearing wear, misalignment, and other mechanical anomalies by monitoring changes in vibration signatures over time.

Such applications emphasize sufficient bandwidth for vibration analysis, low noise for clean frequency spectra, and mechanical robustness to withstand continuous vibration. Early detection of abnormal vibration patterns improves reliability and reduces maintenance costs.

These applications require sufficient bandwidth to capture meaningful vibration content, low noise for clear signal interpretation, and mechanical robustness for continuous operation. The sensor must also support reliable long-term monitoring as vibration amplitude and spectral content evolve with developing faults.

As illustrated in **Figure 6.8**, accelerometers mounted near electric vehicle motors are used to monitor vibration signatures associated with normal operation and emerging mechanical faults. Changes in vibration amplitude, consistency, or pattern over time can indicate imbalance, bearing wear, misalignment, or drivetrain degradation.

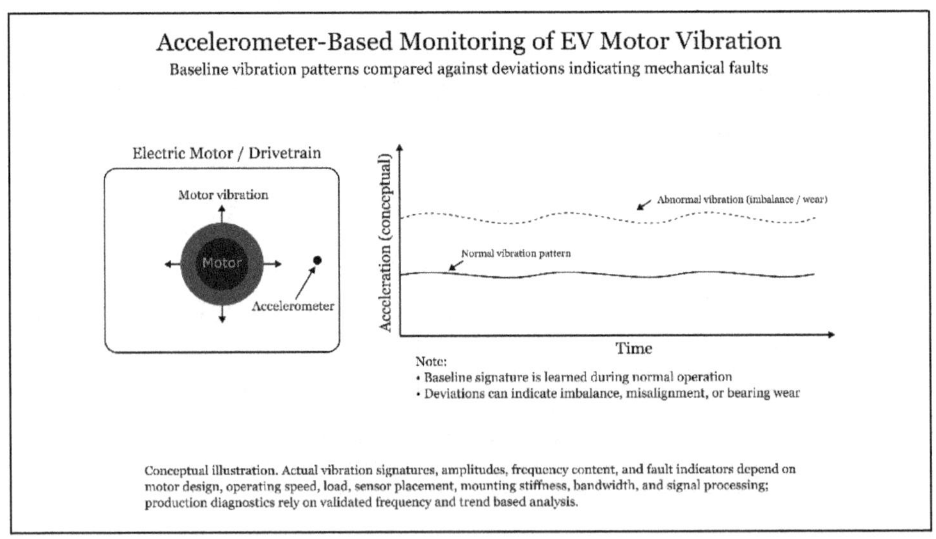

Figure 6.8 Accelerometer-based vibration monitoring of an electric vehicle motor.

6.2.6 Vehicle Security, Impact, and Tilt Detection

Accelerometers are commonly used in vehicle security systems to detect impacts, towing, forced entry, or tilting during wheel theft attempts. These applications rely on accurate gravity-based tilt sensing, programmable thresholds, and reliable wake-on-motion operation.

Low power consumption is essential because security systems often operate continuously while the vehicle is parked. Stable zero-g offset helps prevent false alarms caused by temperature changes or long-term mechanical stress.

Overall, automotive applications show how safety, durability, and long-term reliability shape accelerometer selection and system design. In these environments, parameters such as offset drift, noise, bandwidth, shock survivability, and temperature stability directly affect both system performance and validation requirements.

6.3 Medical and Healthcare Devices

Accelerometers play an increasingly important role in modern medical and healthcare devices. Their ability to sense subtle motion, changes in posture, and vibration makes them well suited for diagnostics, long-term monitoring, rehabilitation, and patient safety. Unlike many medical sensors that require direct physiological contact or complex interfaces, accelerometers can extract clinically useful information from motion alone, enabling compact, wearable, and low-power solutions.

As illustrated in **Figure 6.9**, wearable medical devices use accelerometers to capture raw motion signals that are then interpreted to extract clinically meaningful information.

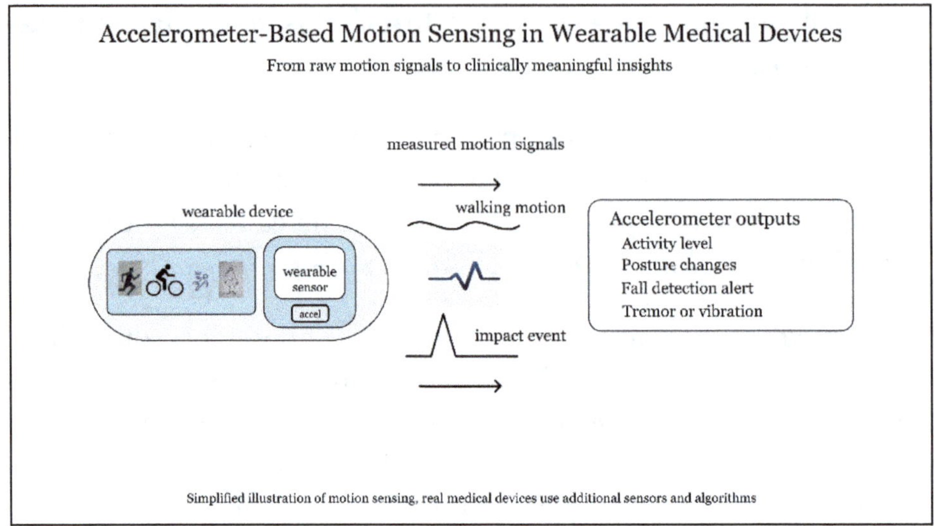

Figure 6.9 Overview of accelerometer-based motion sensing in wearable medical devices.

Characteristic motion patterns associated with walking, posture changes, impacts, or small vibrations can be analyzed to estimate activity level and monitor general movement behavior. They can also be used to detect falls and identify tremors.

While the accelerometer provides the fundamental motion signal, practical medical devices rarely rely on that signal alone. In many cases, the data is combined with additional sensors and algorithms to improve robustness, reduce false detections, and support long-term monitoring in real-world conditions.

Medical applications place distinct demands on accelerometers. Sensors must be small and lightweight for long-term wear, consume minimal power to support continuous monitoring, and remain stable across body temperature ranges and daily activity. Because medical decisions may be influenced by sensor output, consistency, repeatability, and reliability are critical.

6.3.1 Fall Detection and Emergency Alert Systems

Fall detection is one of the most widely deployed medical uses of accelerometers, particularly for elderly individuals and patients with mobility impairments. As shown earlier in **Figure 6.3**, a fall typically produces a recognizable sequence that includes a brief free-fall interval, a sharp impact, and a period of reduced motion afterward. In medical and emergency alert systems, the challenge is not only recognizing this pattern, but doing so reliably across different users, body placements, and everyday non-fall activities.

Reliable detection requires capturing both low-*g* and high-*g* events. This places emphasis on sufficient full-scale range, accurate threshold detection, rapid response, and strong shock resistance. Wake-on-motion capability and efficient interrupt handling are also important to minimize power consumption during continuous monitoring.

6.3.2 Sleep Monitoring and Circadian Rhythm Analysis

Medical-grade sleep monitoring devices use accelerometers to detect nocturnal movement, restlessness, and sleep transitions. Very small motion signals are correlated with sleep stages and sleep quality over long periods.

These applications demand excellent low frequency noise performance, stable offset behavior, and minimal temperature drift. Because monitoring often occurs overnight or continuously for days, low ODR and low power consumption are essential.

6.3.3 Activity Monitoring and Gait Analysis

Accelerometers are widely used to monitor physical activity and analyze gait patterns. Walking, running, and postural transitions generate characteristic low-frequency acceleration signatures that can be quantified to assess mobility, balance, and rehabilitation progress.

As shown in **Figure 6.10**, walking produces a repeatable, low frequency acceleration waveform associated with the gait cycle. Characteristic patterns emerge across successive steps, reflecting the timing and coordination of key walking phases. By analyzing the shape, timing, and consistency of these patterns over time, accelerometer-based systems can extract clinically relevant gait metrics such as stride regularity, symmetry, and variability. Reliable gait analysis depends on low noise, stable sensitivity, appropriate bandwidth for human motion, and accurate multi-axis measurement during continuous wear.

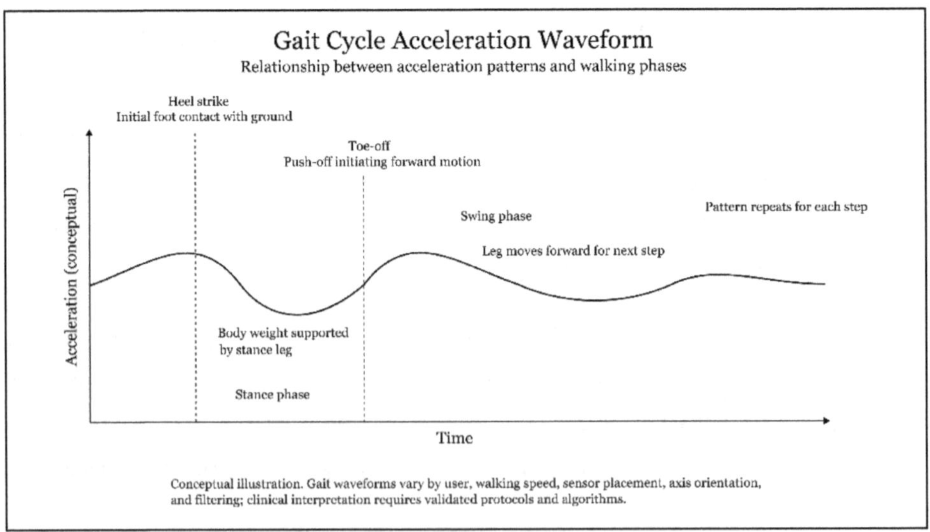

Figure 6.10 Conceptual acceleration waveform corresponding to a single gait cycle.

In clinical settings, gait metrics such as stride variability, symmetry, and walking speed provide insight into conditions including Parkinson's disease, stroke recovery, and post-surgical rehabilitation. These measurements help translate accelerometer data into clinically meaningful indicators of mobility, balance, and recovery progress.

6.3.4 Respiratory and Cardiac Micro-Movement Sensing

Accelerometers are sensitive enough to detect extremely small movements associated with respiration and cardiac activity. Chest expansion during breathing and subtle vibrations related to heart motion can be measured using carefully tuned sensors and signal processing.

These applications rely on ultra-low noise, high resolution, and excellent low frequency stability. Long-term monitoring also requires minimal drift and low power operation. Such techniques are increasingly used in smart patches, wearable monitors, and contactless health-tracking systems.

6.3.5 Tremor Detection and Movement Disorders

Accelerometers are used to quantify tremors and involuntary movement associated with neurological disorders such as Parkinson's disease and essential tremor. Tremors are small, involuntary, rhythmic body movements, often affecting the hands or arms. Tremors typically occur within narrow frequency bands and may have small amplitudes that change over time.

As shown in **Figure 6.11**, tremor-related motion produces a small-amplitude, relatively high-frequency acceleration waveform that is distinct from normal voluntary movement. Unlike gait or posture transitions, tremors are characterized by narrow-band oscillations that persist over time and may vary gradually in amplitude or frequency.

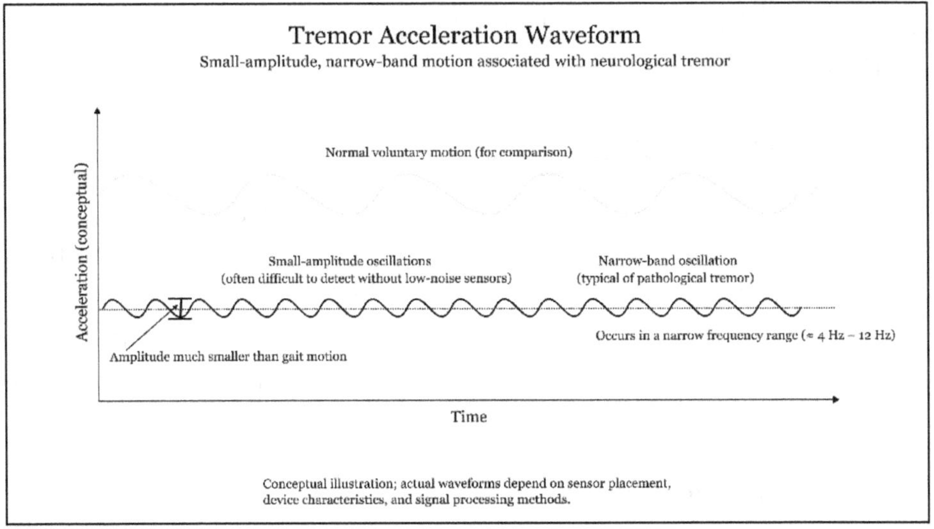

Figure 6.11 Conceptual acceleration waveform associated with neurological tremor.

Accurate tremor detection and tracking require low noise density, adequate resolution, and stable multi-axis measurement so that subtle changes can be quantified reliably. These measurements support objective assessment of movement disorders and help clinicians evaluate disease progression and treatment response.

6.3.6 Rehabilitation and Physical Therapy Devices

In rehabilitation settings, accelerometers track movement quality, range of motion, and exercise compliance. Sensors placed on limbs or joints help quantify smoothness, symmetry, and progress during therapy.

As shown in **Figure 6.12**, accelerometers used in rehabilitation and clinical monitoring are commonly placed at multiple locations on the body, including the wrist, chest, lumbar region, thigh, and lower leg. Each placement captures different aspects of human movement, such as upper-limb activity, posture and balance, gait stability, step timing, and foot impact.

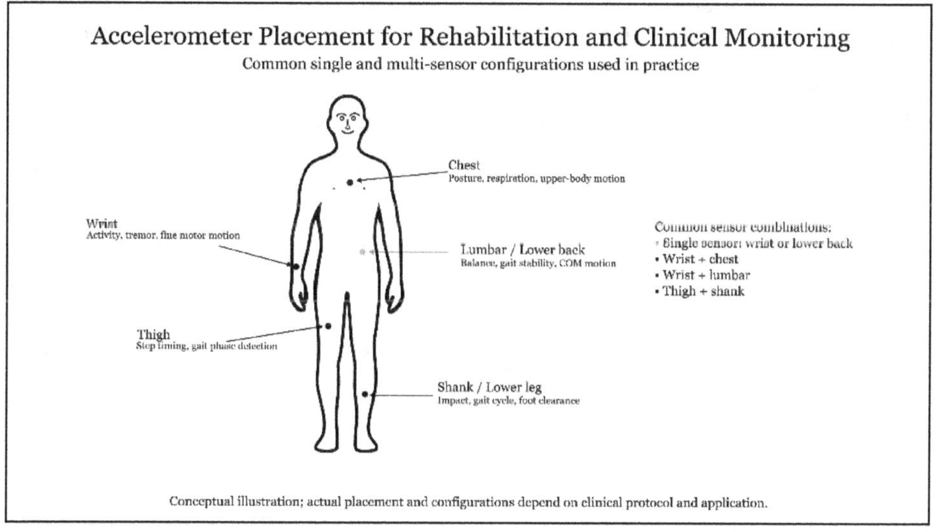

Figure 6.12 Common accelerometer placements for rehabilitation and clinical monitoring.

Depending on the application, systems may use a single sensor (for example, at the wrist or lower back) or combine multiple sensors to improve measurement fidelity and interpret complex motion patterns.

By analyzing multi-axis acceleration across these placements over time, clinicians and rehabilitation systems can quantify range of motion, movement quality, symmetry, and exercise compliance, supporting objective assessment of progress in both in-clinic and remote therapy settings. These systems prioritize multi-axis accuracy, consistent sensitivity, and sufficient dynamic range to capture both slow controlled movements and faster transitions.

6.3.7 Medication Adherence and Smart Drug Delivery

Some connected medical devices use accelerometers to recognize handling and usage patterns. Examples include detecting whether an inhaler was shaken before use, whether an injection pen was oriented correctly during administration, or whether a device experienced a characteristic actuation sequence. These motion signatures can help verify proper technique and support adherence tracking.

Such applications place emphasis on reliable motion-pattern recognition, stable orientation sensing, and low power operation for extended monitoring. In practice, accelerometer data may also be combined with timing logic, device-state information, or other sensors to improve confidence in usage-event detection.

Across medical and healthcare applications, accelerometers demonstrate the ability to detect subtle motion with high reliability over extended periods. In these applications, low noise enables meaningful signal extraction, long-term stability supports continuous monitoring, and low power consumption helps ensure patient comfort and practical use. These requirements differ significantly from those of consumer or automotive systems and reinforce the importance of application-driven sensor selection.

6.4 Industrial and Manufacturing Equipment

Industrial and manufacturing environments place some of the most demanding requirements on accelerometers. Machines often operate continuously, generate significant vibration, and are exposed to temperature variation, mechanical shock, and electrical noise. In these settings, accelerometers are used primarily for diagnostics, fault detection, safety, and predictive maintenance rather than user interaction. The information they provide helps prevent unplanned downtime, reduce maintenance costs, and extend equipment life.

Industrial applications emphasize bandwidth, mechanical robustness, long-term stability, and predictable noise behavior. While many systems use capacitive MEMS accelerometers, higher full-scale ranges or specialized packaging are often selected when vibration levels or shock exposure are severe. Balancing sensitivity with durability is a central design consideration.

6.4.1 Vibration Monitoring and Predictive Maintenance

Beyond individual motor diagnostics, one of the most important industrial uses of accelerometers is vibration monitoring across rotating equipment. Rotating equipment such as motors, pumps, compressors, fans, conveyors, and gearboxes produces characteristic vibration signatures during normal operation. As components wear, loosen, or become misaligned, these signatures change in measurable ways.

As shown in **Figure 6.13**, rotating machinery in normal operation produces relatively stable and repeatable vibration patterns over time. As mechanical faults develop, such as bearing wear, imbalance, misalignment, or looseness, these patterns change in measurable ways, including increases in amplitude, shifts in frequency content, and additional irregular components. By monitoring these changes and tracking trends over time, accelerometer-based condition-monitoring systems can identify early signs of degradation before catastrophic failure occurs.

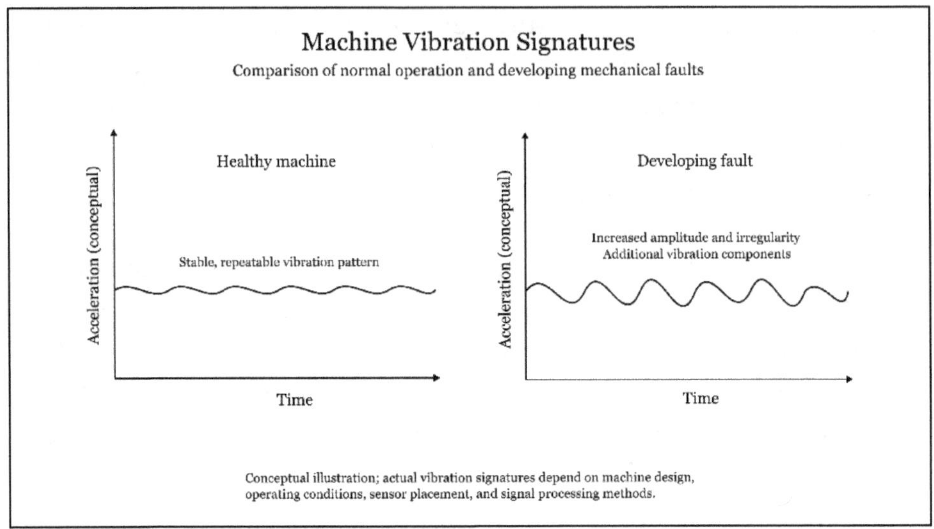

Figure 6.13 Conceptual comparison of vibration signatures from healthy and failing machinery.

Effective vibration monitoring depends on sufficient bandwidth to capture relevant mechanical frequencies, low noise density for sensitivity to subtle changes and spectral clarity, stable sensor behavior over temperature, robust mounting to ensure measurement repeatability, and strong shock resistance to survive abnormal events. By analyzing vibration data over time, accelerometers enable early fault detection, where increases in vibration amplitude at rotational frequencies, the appearance of new harmonics, or elevated broadband noise can indicate bearing degradation, imbalance, or structural looseness.

6.4.2 Motor Health and Machinery Diagnostics

Accelerometers mounted on motor housings or mechanical frames monitor imbalance, shaft misalignment, bearing wear, lubrication issues, and resonance effects. Even small increases in vibration amplitude can signal early-stage faults long before catastrophic failure occurs.

These systems emphasize moderate to high bandwidth, consistent sensitivity across temperature, low noise for clean frequency-domain analysis, and stable mechanical mounting. Sensor behavior under package stress is especially important, as mounting conditions directly influence measurement accuracy.

6.4.3 Robotics and Industrial Automation

Industrial robots and automated manufacturing systems use accelerometers to detect collisions, monitor vibration, and improve motion precision. Accelerometers help identify abnormal forces, resonance during high-speed movement, and mechanical wear in joints or end effectors.

These applications require sufficient bandwidth to capture rapid acceleration changes, low latency for real-time response, synchronized multi-axis measurement, and tolerance to continuous vibration. In collaborative robots, accelerometers also support safety functions by detecting unintended contact with operators.

6.4.4 Heavy Machinery and Mobile Equipment

Construction, mining, and agricultural machinery generate strong vibration and repeated mechanical shock. Accelerometers monitor engine behavior, hydraulic systems, chassis motion, and impact events during operation.

Sensors in these environments must withstand very high shock levels, operate across wide temperature ranges, and maintain stable performance despite harsh conditions. Robust packaging and predictable drift behavior are critical for long-term deployment.

6.4.5 Structural Health Monitoring

Accelerometers are widely used to monitor the health of large structures such as bridges, buildings, towers, wind turbines, and pipelines. These structures experience oscillations due to wind, traffic, thermal expansion, and seismic activity. Changes in vibration patterns or modal frequencies can indicate structural degradation or damage.

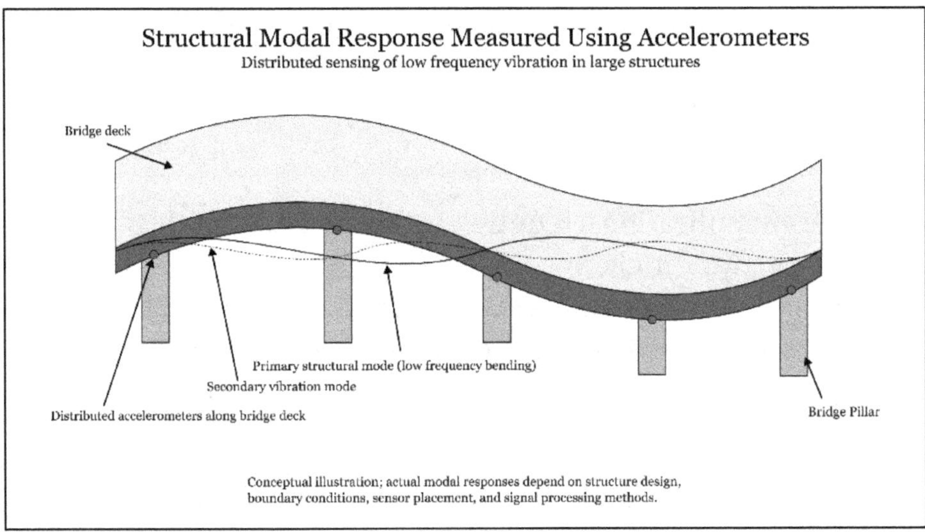

Figure 6.14 Structural modal response measured using distributed accelerometers.

As illustrated in **Figure 6.14** accelerometers distributed along large structures such as bridges enable measurement of low frequency structural vibration and spatial deformation associated with dominant vibration modes.

By capturing acceleration at multiple locations along the bridge deck, the system can reconstruct modal shapes, identify resonance behavior, and track changes in modal characteristics over time. Differences between primary and secondary mode shapes reflect how the structure responds to environmental excitation, traffic loading, wind, and thermal effects.

Long-term shifts in modal frequency (a characteristic natural vibration frequency of the structure), damping, or spatial response can indicate structural degradation, damage, or changes in boundary conditions. Structural health monitoring therefore prioritizes very low noise performance to detect small-amplitude motion, adequate low-frequency bandwidth, stable sensitivity, long-term offset stability, and minimal temperature drift.

Because measurements often span years, consistency and repeatability over time are more important than short-term precision, enabling reliable modal identification and trend analysis throughout the structure's operational life.

6.4.6 Conveyors, Production Lines, and Assembly Systems

Manufacturing lines use accelerometers to detect misalignment, roller wear, jams, and imbalance in conveyors and automated equipment. Increasing vibration often precedes mechanical failure, allowing maintenance to be scheduled proactively. These systems emphasize bandwidth appropriate to mechanical speeds, low noise for clear signal interpretation, and high durability under continuous operation.

Across these applications, industrial and manufacturing systems highlight the accelerometer's role as a diagnostic and monitoring tool rather than a user interface sensor. In these systems, bandwidth determines which vibration signatures can be detected, noise limits sensitivity to early faults, stability supports long-term trend analysis, and shock resistance ensures survivability under abnormal conditions. These requirements demonstrate how accelerometer specifications translate directly into operational reliability and cost savings.

6.5 Aerospace, Drones, and Robotics

Aerospace systems, drones, and robotic platforms rely heavily on accelerometers for control, navigation, and stabilization. In these applications, accelerometers are not auxiliary sensors but core components of motion control systems. Accurate acceleration measurement is essential for maintaining balance, responding to disturbances, and executing precise movements in real time.

Compared to consumer devices, these systems operate under tighter latency constraints and often experience continuous vibration, rapid dynamic motion, and changing environmental conditions. Accelerometers used here must deliver consistent multi-axis measurements with predictable behavior, low latency, and sufficient bandwidth to support fast control loops.

6.5.1 Drone Stabilization and Flight Control

In multirotor drones, accelerometers measure translational acceleration and operate alongside gyroscopes to estimate orientation changes, velocity, and external disturbances such as wind gusts. This information feeds into flight control algorithms that continuously adjust motor speeds to maintain stable flight.

As shown in **Figure 6.15**, accelerometers mounted in a drone measure the projection of the gravity vector onto the body-fixed axes, enabling estimation of pitch and roll orientation. When the drone tilts, the gravity vector resolves into components along the lateral and longitudinal axes, providing attitude information that complements gyroscope measurements. In steady flight or hovering conditions, accurate gravity estimation supports stable attitude control and drift correction. During dynamic maneuvers, accelerometer data also reflects translational acceleration and external disturbances, requiring sensor fusion with gyroscopes to separate motion-induced acceleration from gravity.

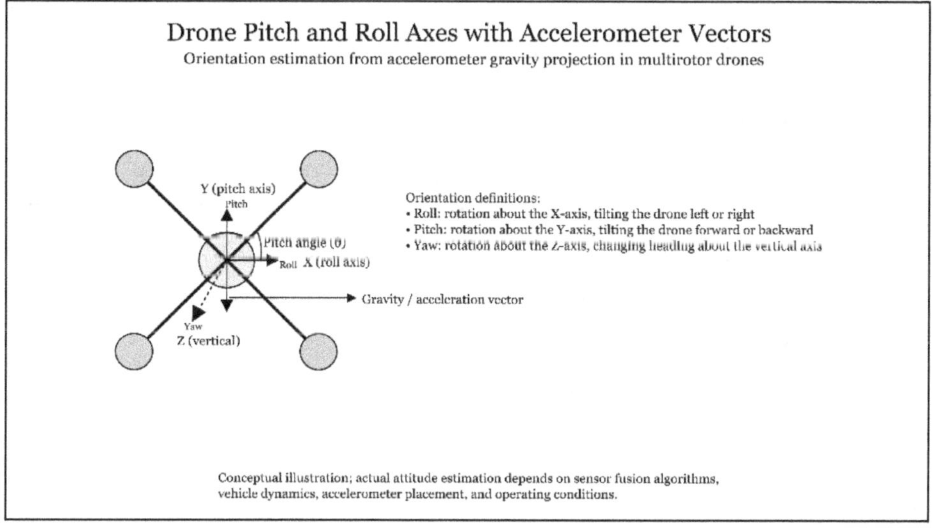

Figure 6.15 Drone pitch and roll axes with accelerometer-derived gravity vector.

Hovering and slow maneuvers emphasize low noise and stable offset behavior, since small errors can accumulate into visible drift or oscillation. Faster maneuvers and disturbance rejection require sufficient bandwidth and high ODR to capture rapid acceleration changes. Low latency and synchronized sampling with gyroscopes are essential for tight control loops and predictable response. These measurements are central to flight control algorithms that maintain stability, reject disturbances, and execute controlled maneuvers.

6.5.2 Robotics: Motion Control and Vibration Compensation

Robotic systems use accelerometers to monitor motion quality, detect unexpected forces, and compensate for vibration. In industrial robots, accelerometers help identify resonance in joints, overshoot during high-speed motion, and vibration that reduces positioning accuracy.

Collaborative robots also rely on accelerometers as part of their safety systems. Sudden changes in acceleration can indicate contact with a human or an unexpected obstacle, triggering force limiting or shutdown mechanisms. These applications require fast response, reliable threshold detection, and predictable behavior under mechanical stress.

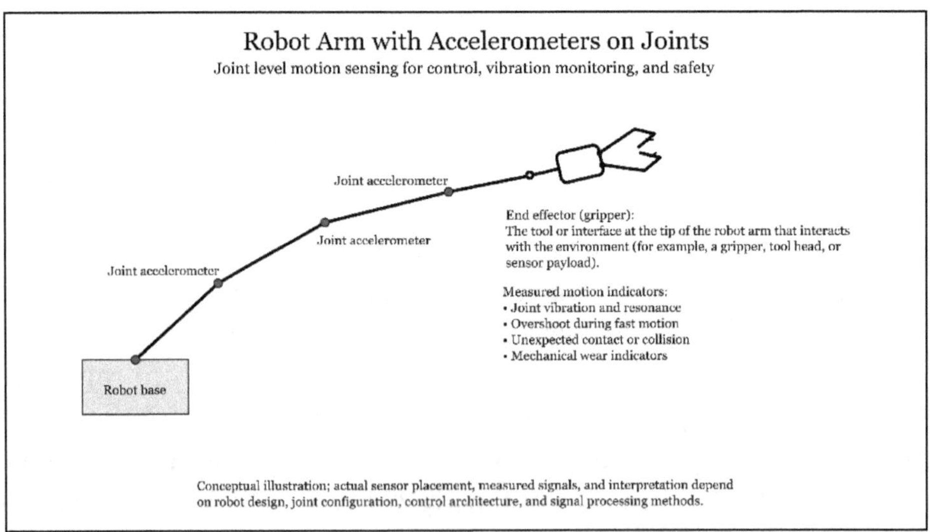

Figure 6.16 Robot arm with accelerometers mounted at joints.

As shown in **Figure 6.16**, accelerometers mounted near robotic joints measure local acceleration associated with joint motion, vibration, and dynamic loading. By monitoring acceleration at multiple joints, robotic control systems can detect resonance, overshoot (motion that briefly exceeds the intended position or response), mechanical wear, and unexpected contact events.

Together, these measurements support vibration compensation, collision detection, safety functions, and motion quality monitoring in robotic systems. To be effective, accelerometer data must be delivered with low latency, stable sensitivity, and predictable behavior under continuous mechanical stress.

In mobile robots and automated guided vehicles, accelerometers support local motion estimation, terrain assessment, and vibration monitoring on uneven surfaces. Stable multi-axis measurement and low latency are critical for maintaining accurate motion estimates.

6.5.3 Aerospace and Aviation Systems

Accelerometers play important roles in both aircraft and spacecraft, although performance requirements vary by application. In aircraft, accelerometers contribute to attitude estimation by providing gravity-based reference information, and are also used for turbulence detection, structural vibration monitoring, and navigation backup when satellite positioning is unavailable.

In spacecraft and satellites, accelerometers measure small non-gravitational accelerations in near-zero-gravity conditions, including disturbances from atmospheric drag (very thin residual air resistance in low Earth orbit), onboard activity, and transient acceleration caused by thruster firings, as well as structural vibration during launch and operation. These environments impose stringent demands on stability, temperature behavior, and mechanical robustness.

While mission-critical aerospace guidance systems often use specialized high-grade sensors, MEMS accelerometers are widely employed in unmanned aerial vehicles, experimental platforms, payload instrumentation, and research systems where size, weight, and power consumption are constrained.

6.5.4 Autonomous Navigation and Inertial Sensing

Accelerometers form a core part of inertial measurement units used in autonomous aerial and robotic systems. When combined with gyroscopes, they support short-term dead reckoning, motion estimation during GPS or Global Navigation Satellite System (GNSS) outages, and detection of slippage (loss of traction causing actual motion to differ from expected motion) or uneven terrain.

These applications emphasize low noise, stable bias behavior, accurate axis alignment, and fast startup. Although accelerometer errors accumulate over time, well-characterized performance enables reliable navigation over short durations and supports higher-level sensor fusion.

6.5.5 Gimbals, Stabilized Cameras, and Precision Payloads

Accelerometers are used in camera gimbals and stabilized payloads to detect translational motion and vibration. When paired with gyroscopes, they help maintain level orientation and smooth motion during flight or movement.

These systems require low noise, good linearity, stable zero-g offset, and high ODR to detect fast disturbances. In mapping and photogrammetry applications, improved accelerometer accuracy contributes to better image alignment and data quality.

Collectively, aerospace, drone, and robotic applications highlight the accelerometer's role in real-time control systems where latency, stability, and predictability are critical. Noise influences control smoothness, bandwidth limits disturbance rejection, offset drift affects long-term stability, and mechanical robustness ensures survivability in dynamic environments. These systems show how accelerometer specifications translate directly into system-level performance.

6.6 Takeaways

In this chapter, we explored how accelerometers shape modern technology across a wide range of applications and real-world use cases. While the internal operation and specifications of accelerometers define what they are capable of, it is within complete systems that these capabilities become meaningful. Consumer electronics, vehicles, medical devices, industrial machinery, aerospace platforms, and everyday embedded products all rely on accelerometers to translate physical motion into actionable information.

Across these domains, accelerometers are used not simply to record acceleration, but to interpret motion patterns, classify events, stabilize systems, and enable intelligent behavior. In smartphones and wearables, they support functions such as orientation detection, gesture recognition, image stabilization, and activity tracking. In vehicles, they contribute to safety-critical functions such as airbag deployment, stability control, and rollover detection. In medical devices, they help monitor falls, gait, tremors, sleep, and subtle physiological motion. Industrial systems rely on them for vibration analysis, fault detection, and predictive maintenance. In robotics, drones, and aerospace platforms, accelerometers form a core part of inertial sensing for balance, navigation, and precise motion control. Even household and embedded products depend on them to detect imbalance, movement, or impacts.

These examples reinforce a key theme developed throughout this book: accelerometer specifications matter differently depending on application context. Wearable devices prioritize low power consumption and low noise. Automotive crash sensors demand fast response and high shock tolerance. Industrial monitoring systems emphasize bandwidth and long-term stability. Aerospace, drones, and robotic platforms depend on low latency, accurate multi-axis alignment, and predictable behavior under dynamic conditions. The application ultimately determines which performance parameters are critical and which trade-offs are acceptable.

These use cases also show that raw accelerometer outputs become meaningful only through interpretation. Engineers apply filtering, thresholds, signal processing, and system-level logic to extract higher-level information such as events, motion states, or fault conditions. Accelerometers rarely operate in isolation; instead, they function as part of integrated sensing and decision-making systems that may include additional sensors, timing logic, and contextual awareness.

With a clear understanding of how accelerometers are applied in real products, we are now positioned to look ahead. In the next chapter, we will explore current and emerging trends in accelerometer technology, including advances in ultra-low-power design, sensor fusion, intelligent signal processing, and new applications shaping the next generation of motion-sensing systems.

CHAPTER 7
Current and Emerging Trends

Accelerometer technology continues to evolve as advances in manufacturing, signal processing, and system integration push the boundaries of what these sensors can deliver. While the fundamental physics of capacitive MEMS accelerometers remain unchanged, improvements in noise performance, power efficiency, on-chip processing, and packaging are reshaping how accelerometers are used in modern devices. These developments do not replace the principles discussed in earlier chapters; instead, they refine them, making accelerometers smaller, more efficient, more accurate, and better suited for intelligent systems.

In this chapter, we focus on realistic and technically grounded trends that engineers are already encountering or can expect to see in the near future. These include the continued push toward ultra-low-power operation, the growing role of intelligent on-sensor processing, deeper integration with sensor fusion and wireless systems, ongoing miniaturization, and the expansion of accelerometer use across medical, industrial, and consumer applications. The goal is not to predict distant or speculative futures, but to highlight practical developments that are shaping the next generation of motion sensing.

7.1 Ultra-Low Power Motion Sensors

One of the most significant trends in modern accelerometer design is the continued push toward ultra-low power operation. As wearables, fitness trackers, medical patches, IoT nodes, and battery-powered industrial sensors become more common, power consumption has become one of the most important differentiators in sensor selection. Manufacturers are increasingly optimizing MEMS structures and ASIC circuitry to reduce current consumption without sacrificing essential performance.

Ultra-low power accelerometers still rely on the same capacitive MEMS principles described earlier in this book, but benefit from substantial improvements in circuit efficiency, signal-chain architecture, and power management. Refinements in charge-balancing techniques, switched-capacitor readout circuits, and low-leakage design have reduced quiescent current from tens of microamps to single-digit microamps, and in some operating modes to the sub-microamp range.

7.1.1 Smarter Power Modes

Early accelerometers typically supported only active and standby modes. Modern devices now offer a range of power states that allow fine-grained control over energy consumption. These include multiple low power measurement modes, wake-on-motion operation consuming well below one microamp, dynamic mode switching based on detected activity, and event-driven sampling that replaces continuous high-rate acquisition.

As a result, sensors can remain responsive to motion while consuming negligible energy during periods of inactivity. This capability is especially important in devices that must operate continuously for days, weeks, or longer on a limited power source.

The typical operating current ranges illustrate this trend clearly. A single device may support an active mode in the 10 μA to 20 μA range, a low-power measurement mode at a few microamps, wake-on-motion below one microamp, and deep-sleep currents on the order of hundreds of nanoamps.

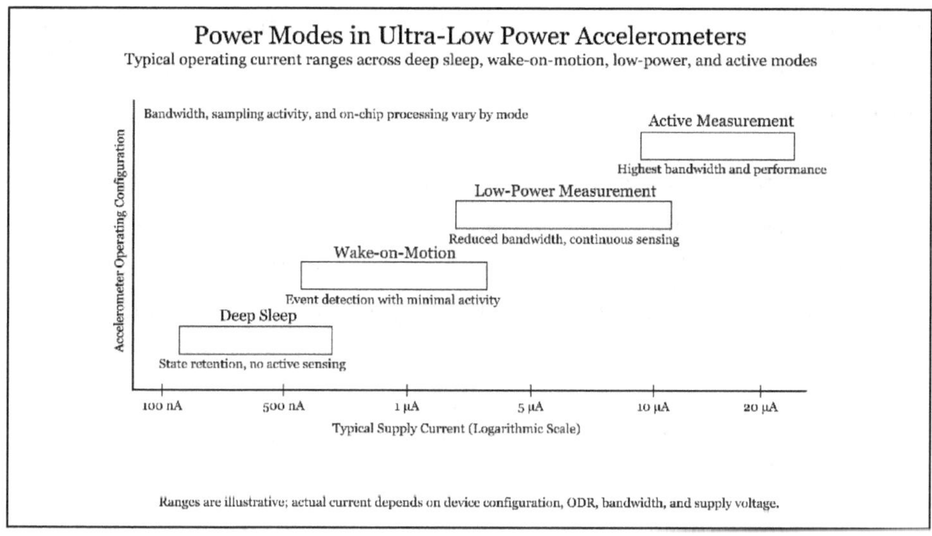

Figure 7.1 Power Modes Comparison in Ultra-Low-Power Accelerometers.

Figure 7.1 illustrates the wide dynamic range of power states supported by modern ultra-low-power accelerometers. Rather than operating in a single continuous mode, these devices offer multiple operating states optimized for responsiveness, measurement fidelity, or energy conservation. Deep sleep modes minimize current draw while retaining configuration state, wake-on-motion modes enable event detection at sub-microamp levels, and low-power measurement modes support continuous sensing with reduced bandwidth. Full active modes provide maximum performance at higher current levels. The ability to transition dynamically between these modes is a key enabler of long battery life in always-on and battery-constrained systems.

7.1.2 Integrated Event Detection to Reduce System Power

To further minimize overall power consumption, accelerometers increasingly include integrated event detection capabilities.

Common examples include step detection, tap or double-tap detection, freefall detection, tilt detection, activity or inactivity classification, and programmable motion thresholds.

By identifying events internally, the accelerometer wakes the host microcontroller only when meaningful motion occurs. This approach dramatically reduces the time the host processor must remain active and often provides larger energy savings than optimizing the sensor alone.

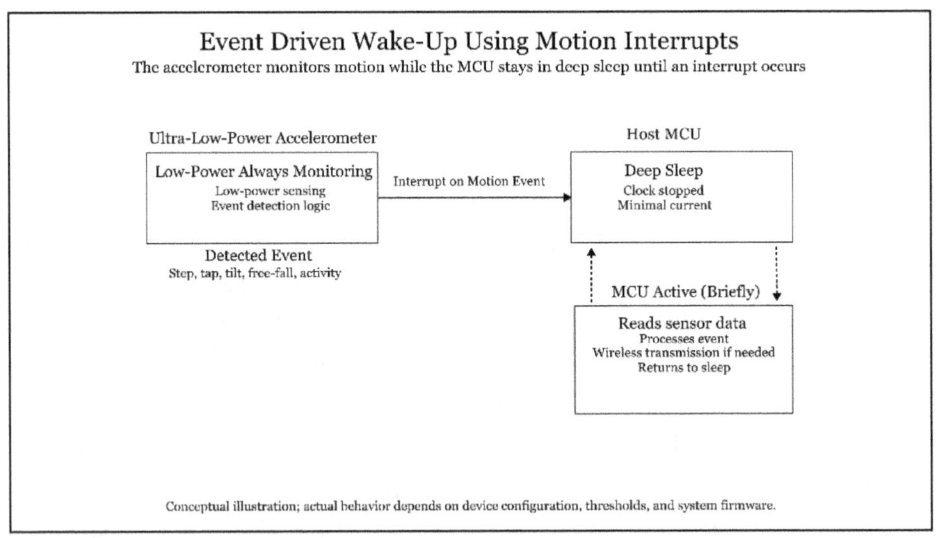

Figure 7.2 Event-Driven Wake-Up Using Integrated Motion Detection.

Figure 7.2 illustrates an event-driven sensing architecture in which the accelerometer remains active at ultra-low power while the host microcontroller stays in deep sleep. The sensor continuously monitors motion and evaluates programmable conditions such as thresholds, activity classification, or specific events. When a relevant event is detected, the accelerometer asserts an interrupt that briefly wakes the microcontroller to process the event, log data, or transmit information wirelessly. After handling the event, the microcontroller returns to sleep.

By avoiding continuous MCU operation and high-rate polling, this architecture often yields greater energy savings at the system level than reducing sensor current alone. This trend is already well established in wearables and medical devices and is expanding rapidly into industrial IoT systems, where long battery life is a primary requirement.

7.1.3 Duty-Cycling and Context-Aware Sampling

Duty cycling has become a key technique in reducing average power consumption. Instead of operating continuously, the accelerometer alternates rapidly between active and sleep states while maintaining sufficient responsiveness. Because many motion signals of interest, such as walking, posture, or tilt, are low frequency in nature, intermittent sampling can capture the required information without loss of fidelity.

Some devices extend this concept further by incorporating context-aware logic that adapts sampling behavior based on operating conditions. Sampling rates and bandwidth may change depending on whether the system is stationary, experiencing light motion, or undergoing rapid dynamics. This prevents unnecessary high-rate sampling when it is not required and conserves energy at the system level.

7.1.4 Applications Driving Ultra-Low Power Development

Several application areas are driving the rapid adoption of ultra-low-power accelerometers. Wearables and health monitoring devices rely on continuous motion monitoring, where battery life directly affects user acceptance. Medical patches and biosensors require long-term operation with minimal maintenance while maintaining low noise and stable performance. Asset tracking and IoT nodes use motion detection to reduce wireless transmissions, significantly lowering energy consumption. Industrial wireless sensors benefit from low-power operation that enables battery-powered or energy-harvesting deployments.

Future improvements in ultra-low-power accelerometers will continue to be incremental rather than disruptive. Expected advances include more efficient MEMS structures that require smaller displacements for sensing, further optimization of charge-amplifier and ADC architectures, smarter on-chip decision logic, and dynamic scaling of analog and digital performance based on operating conditions. Tighter integration with microcontrollers, wake-up radios, and low-power wireless systems will further reduce system-level energy consumption.

7.2 AI-Enabled Motion Detection

A growing trend is the use of AI-assisted motion interpretation enabled by accelerometer data, either directly on the sensor or within the surrounding system architecture. As wearables, AR and VR devices, smart appliances, industrial monitors, and medical sensors become more capable, simple threshold-based or rule-based detection is often insufficient. Many applications require recognition of complex or variable motion patterns, and compact AI techniques are increasingly used to extract this information reliably.

AI-enabled motion detection does not replace the underlying MEMS sensing principles or the conventional signal chain. Instead, it builds on them by identifying motion sequences that are difficult to classify using fixed thresholds or handcrafted rules alone, improving robustness to variations in users, mounting conditions, and operating environments.

7.2.1 AI for Motion Classification and Noise Robustness

Even when AI is not embedded directly in the sensor, many systems apply small-footprint AI models at the firmware or application level to improve motion interpretation. These models analyze sequences of acceleration data rather than individual samples, which makes them more tolerant of noise, orientation changes, and user variability.

Common applications include step detection that distinguishes real walking from arm motion, gesture recognition in consumer devices, tremor classification in medical monitoring, anomaly detection in industrial vibration data, and fall detection systems that reduce false alarms. By focusing on patterns over time, AI-based approaches outperform static thresholds in dynamic real-world conditions.

7.2.2 On-Sensor Machine Learning

Some modern accelerometers incorporate small machine-learning models directly within the sensor ASIC. These models, often based on compact decision trees or simple neural networks, operate at very low power levels and classify motion events such as steps versus non-steps, taps versus noise, specific gestures, free-fall versus impact, or broader activity states such as walking, running, sitting, or lying down.

This approach is particularly effective in battery-constrained systems, where waking the host microcontroller frequently would dominate power consumption. Because the models are small, require limited memory, and operate at modest sampling rates, on-sensor inference can be performed at micro-watt power levels (typically corresponding to sub-microampere to a few microamperes of current, depending on supply voltage) without compromising responsiveness.

Figure 7.3 illustrates how embedded AI models can be integrated directly into the accelerometer signal chain. Rather than operating on raw acceleration samples, these embedded models process a small set of extracted features derived from conditioned sensor data. Compact decision trees or small neural classifiers evaluate motion patterns and produce high-level classifications such as steps, gestures, or activity states. Because the models are deterministic, use fixed-point arithmetic, and operate at modest sampling rates, inference can be performed at micro-watt power levels.

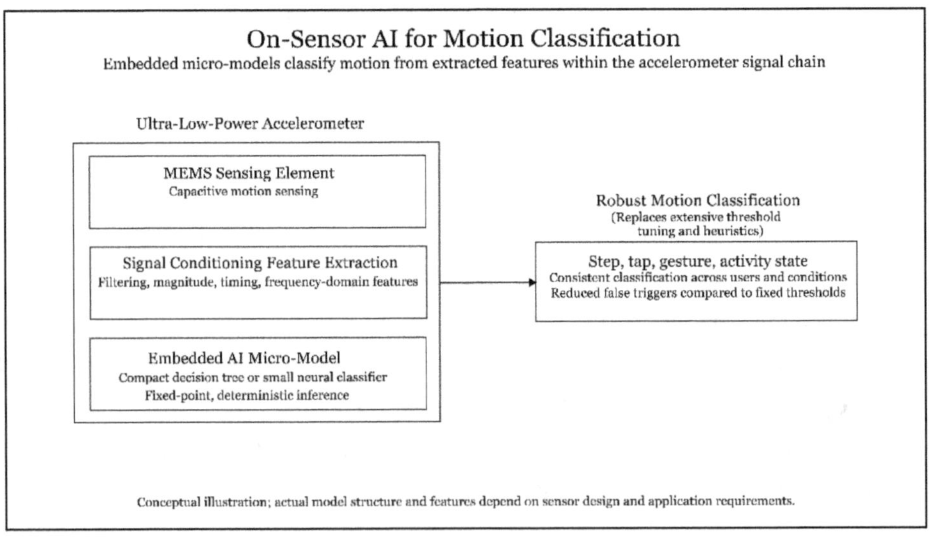

Figure 7.3 Embedded AI Micro-Model for Motion Classification.

7.2.3 Energy-Efficient Models for Edge and On-Sensor AI

AI deployed in accelerometer systems must meet strict constraints. Models must be compact, deterministic, low power, stable across temperature, and simple to update through firmware. This has driven the development of optimized decision trees, small neural networks, sensor-specific classifiers, and inference engines that operate using fixed-point arithmetic.

Unlike cloud-based AI, these models are designed for predictability and efficiency rather than raw computational power. They are practical today and are already integrated into many modern motion sensors and low-power embedded platforms.

7.2.4 Context-Aware Motion Interpretation

A key advantage of AI-based approaches is the ability to incorporate context into motion classification. Instead of evaluating acceleration values in isolation, AI models can account for factors such as recent activity history, time-dependent behavior, known operating states, or expected motion profiles.

For example, an industrial vibration sensor may interpret the same acceleration signature differently depending on whether a machine is starting up, operating under load, idling, or shutting down. Similarly, a wearable device may classify motion differently based on time of day or prior user activity. This contextual awareness improves accuracy and reduces false detections in complex environments.

Figure 7.4 illustrates a context-aware motion classification pipeline in which acceleration data is processed through multiple stages before a final interpretation is produced. Raw sensor samples are first converted into descriptive features, which are then aggregated over time to capture motion sequences rather than isolated events.

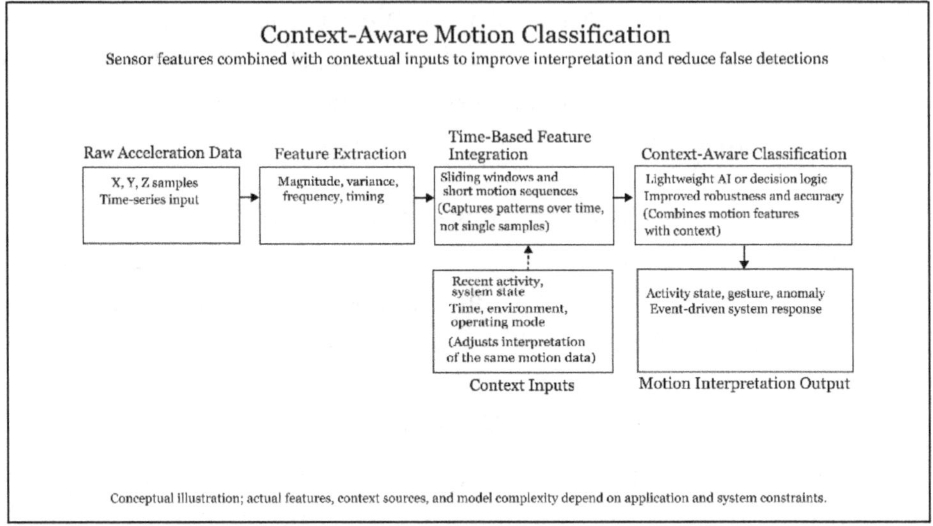

Figure 7.4 Context-Aware Motion Classification Pipeline.

Within this framework, contextual information such as recent activity history, system operating state, or environmental conditions is incorporated alongside these features to guide classification. By evaluating motion patterns within a broader context, simple AI or decision logic can distinguish between similar acceleration signatures and reduce false detections. This approach improves reliability in real-world conditions while remaining compatible with the power and resource constraints of edge and on-sensor systems.

Over the next several product generations, AI-enabled motion detection is expected to become more widespread rather than more complex. Anticipated developments include broader availability of compact on-sensor classification engines, standardized interfaces for AI-ready sensors, dynamically adaptive thresholds, and automatic classification of common motion states such as tilt, tap, vibration mode, or activity level.

These improvements will remain incremental and grounded in existing technology, but they will significantly reduce host processor workload, improve detection accuracy, and enable richer motion-aware applications without increasing system power consumption.

Collectively, AI-enabled accelerometer-based motion-detection systems demonstrate how improved classification accuracy, reduced host workload, and low-power operation can be achieved together in practical edge devices. Compared with fixed thresholds or hand-tuned rule-based logic, these approaches scale more naturally across users, device orientations, and operating conditions. This makes them especially useful in always-on systems where robust classification must be achieved with minimal host involvement and tight energy budgets.

7.3 Integrated Inertial Sensor Architectures

Another important trend in accelerometer development is the increasing integration of sensing and motion-processing functions into compact sensor architectures. Traditionally, accelerometer data, gyroscope data, magnetometer data, and sometimes pressure-sensor data were handled as separate streams and combined in firmware on the host processor. Today, many devices increasingly embed fusion logic within the sensor or its companion ASIC, while others integrate multiple sensing elements into tightly coordinated packages.

This section examines both developments. First, it explains how embedded fusion engines move synchronization, calibration, and motion interpretation closer to the sensing elements, reducing host processing requirements, lowering system power consumption, and improving output consistency. It then examines the broader integration of multiple sensors in compact packages, where tighter physical and electrical coordination helps improve timing alignment, reduce system complexity, and support more advanced motion and orientation functions.

7.3.1 Embedded Sensor Fusion Engines

Accelerometer data alone is often insufficient for determining orientation, heading, or rotational motion under dynamic conditions. Accurate motion tracking therefore requires combining accelerometer measurements with gyroscope measurements of angular velocity and, in some cases, magnetometer data for heading reference. Traditionally, this sensor fusion was implemented in firmware on the host microcontroller, where multiple sensor data streams had to be synchronized, calibrated, and processed before meaningful motion or orientation outputs could be generated.

Implementing fusion at the host level increases software complexity, host wakeups, and tuning effort, and can introduce variability in performance across devices and operating conditions. By moving fusion logic closer to the sensing elements, manufacturers reduce system complexity and improve consistency, latency, and power efficiency.

Figure 7.5 compares traditional host-based fusion with on-sensor fusion architectures. The illustration is conceptual; sensors may be integrated within a single package or implemented as separate devices depending on system design. In some architectures, external sensors such as gyroscopes or magnetometers are connected through a local digital interface, often using an I^2C master function within the sensor hub, IMU, or companion ASIC rather than the host processor.

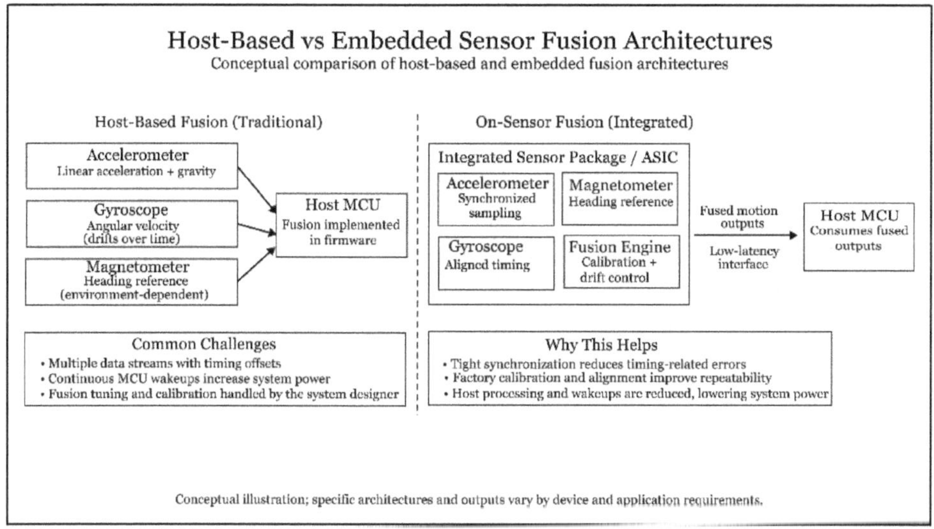

Figure 7.5 Conceptual comparison of host-based versus on-sensor fusion architectures.

In embedded fusion designs, sensor data is combined closer to the point of acquisition, allowing synchronization, calibration, and drift management to be handled more consistently before the data reaches the host. By outputting fused motion information through a low-latency interface, these architectures reduce host processing requirements, lower system power consumption, and improve repeatability across devices and operating conditions.

Many modern inertial measurement units now include dedicated fusion processors that implement these capabilities directly within the device. These embedded fusion engines output information such as orientation quaternions, Euler angles, calibrated acceleration, calibrated angular rate, and basic motion states such as stationary, walking, or rotating.

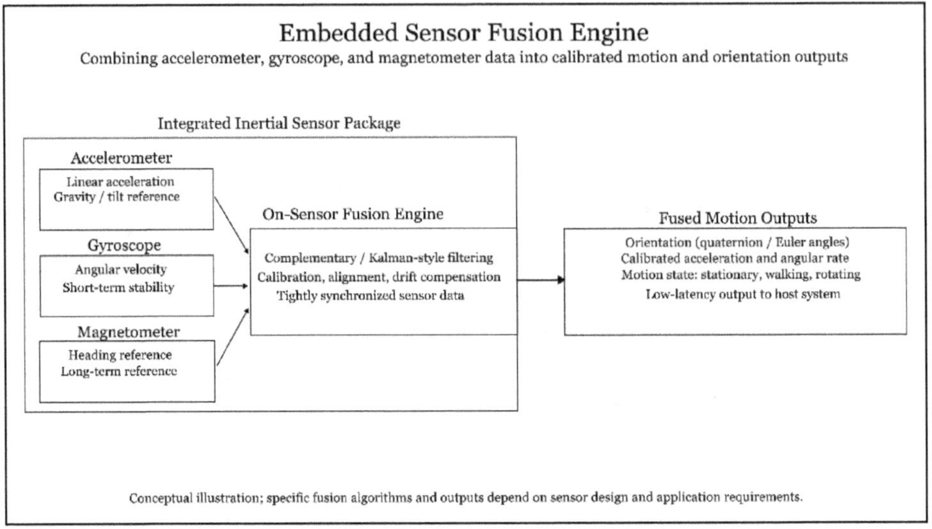

Figure 7.6 Integrated On-Sensor Fusion Engine.

Figure 7.6 illustrates an embedded fusion architecture in which data from multiple inertial sensors are combined directly within the sensor package or its companion ASIC. Accelerometer, gyroscope, and magnetometer measurements are synchronized and processed by a dedicated fusion engine that helps compensate for noise, drift, alignment errors, and timing offsets. The resulting outputs may include calibrated motion signals as well as higher-level motion and orientation information.

By outputting calibrated motion and orientation information rather than raw sensor data, embedded fusion engines reduce host processing workload and simplify system integration for products that cannot support complex host-side fusion algorithms.

These benefits are especially valuable in wearables, AR and VR controllers, robotics platforms, consumer electronics, and camera stabilization systems, where latency, power consumption, and repeatability are critical. Fusion engines are increasingly designed with power-sensitive applications in mind, often supporting partial-update modes and reduced computation when motion is limited.

This approach allows devices such as wearables and IoT nodes to provide reliable orientation and motion classification with less frequent host involvement, improving system efficiency while preserving responsive motion tracking.

7.3.2 Integration of Multiple Sensors in Compact Packages

Future motion-sensing modules are exploring tighter integration of accelerometers, gyroscopes, and related sensing or processing functions within compact packages. In some applications, this may also include additional elements such as magnetometers, pressure sensors, fusion hardware, low-power processing blocks, or dedicated motion engines. The degree of integration depends strongly on application requirements, packaging constraints, and the tradeoffs associated with combining sensors that have different environmental and mounting needs.

Figure 7.7 illustrates the evolution of integrated sensor modules as multiple sensing elements are combined within a single package or closely coordinated module. Early designs commonly paired an accelerometer with a gyroscope, while some later solutions explored the integration of additional sensors such as magnetometers or pressure sensors, depending on the target application and packaging constraints.

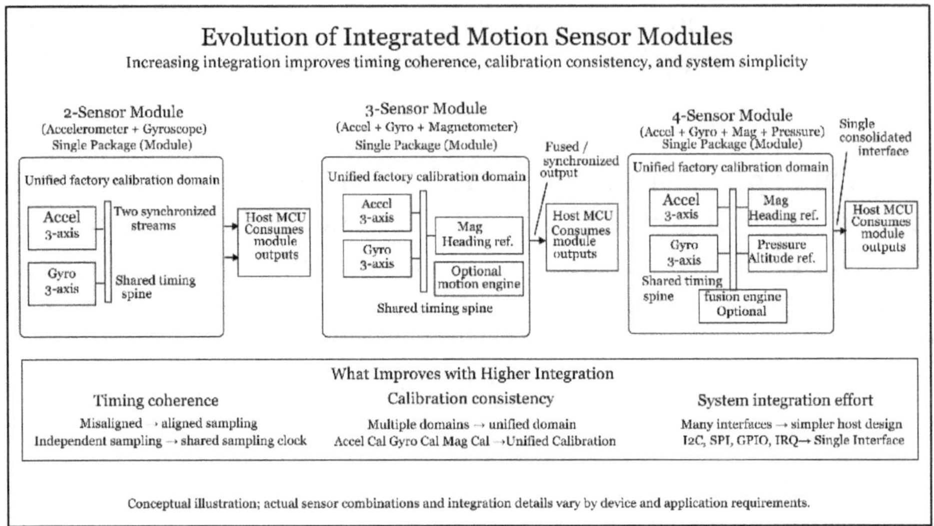

Figure 7.7 Evolution of Integrated Sensor Modules.

As integration increases, sensors share timing, calibration, and processing resources, improving sampling coherence and calibration consistency across sensing axes. Higher integration also reduces system integration effort by simplifying host interfaces and firmware requirements. Together, these improvements enable more accurate and stable system-level sensing while lowering complexity in compact, power-constrained designs.

Integration also reduces bill-of-materials cost and simplifies PCB layout in compact systems. These packaging benefits improve manufacturability while supporting more stable timing, orientation estimation, and motion tracking.

Many integrated sensor packages also incorporate supporting functions beyond the sensing elements themselves, including sensor fusion engines, power-management circuitry, and other processing blocks. This broader functional integration points toward more compact and capable modules that do more than simply combine raw sensor outputs.

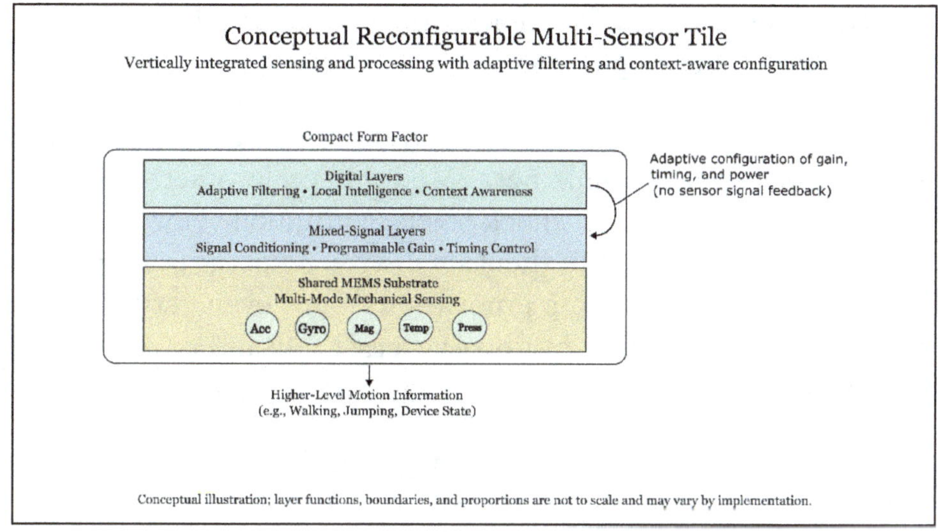

Figure 7.8 Conceptual Reconfigurable Multi-Sensor Tile.

Figure 7.8 illustrates a forward-looking concept in which multiple sensing functions are implemented within a tightly integrated sensor tile rather than as discrete sensors. A shared MEMS substrate supports multi-mode mechanical sensing, while vertically stacked mixed-signal and digital layers enable adaptive filtering, local intelligence, and context-aware configuration of gain, timing, and power.

Rather than relying only on fixed raw sensor streams, the tile or module can produce higher-level motion or device-state information based on system context. This architecture reflects the long-term direction of miniaturization, where tighter integration enables more efficient, robust, and adaptive sensing within extremely compact form factors.

This level of integration also simplifies product design by reducing board-level complexity and enabling more compact system architectures. The trend is especially visible in integrated inertial modules used for smartphones, AR and VR controllers, drones, and robotics platforms.

As packages shrink, manufacturers are simultaneously improving mechanical robustness. Modern accelerometers are designed to withstand high shock levels, continuous vibration, and mechanical stress despite their small size. Structural design improvements and packaging innovations allow even very compact sensors to survive shock events of several thousand g while maintaining stable performance.

Over the next several product generations, motion-sensing architectures are expected to become more tightly integrated, combining sensing elements, calibration resources, fusion logic, and supporting processing functions within increasingly compact modules. More accelerometer families will offer on-sensor fusion outputs, improved calibration and drift compensation, and better coordination with host timing and power-management schemes.

At the same time, future designs may extend beyond traditional inertial combinations to include additional sensing functions, more adaptive processing, and selective use of lightweight machine-learning techniques. These developments will make motion-sensing systems more compact, more power-efficient, and easier to deploy across a wider range of applications without changing the fundamental sensing principles described throughout this book.

7.4 Miniaturization and MEMS Integration

As electronic systems continue to shrink, accelerometer packages are becoming smaller and more tightly integrated. Advances in MEMS fabrication, wafer-level packaging, and mixed-signal ASIC design have enabled accelerometers to fit into increasingly compact footprints while maintaining, and in some cases improving, performance. Miniaturization is not only about saving space; it also improves manufacturability, reduces system cost, and allows sensors to be placed where larger packages would be impractical.

One of the most important developments enabling miniaturization is wafer-level chip-scale packaging (WLCSP). In this approach, the MEMS sensing structure and signal-processing ASIC are integrated and packaged at the wafer level rather than using traditional molded packages or leadframes.

By eliminating leadframes and molded encapsulation, WLCSP reduces package size, assembly complexity, parasitic electrical effects, and mechanical mass while improving manufacturability and performance consistency. This architecture enables extremely compact and robust accelerometers well suited for space-constrained systems.

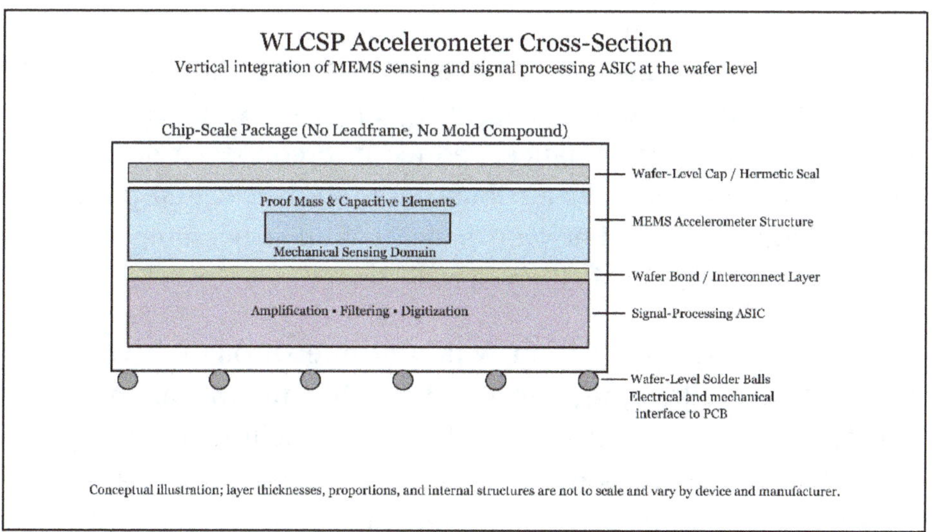

Figure 7.9 WLCSP Accelerometer Cross-Section.

Figure 7.9 Error! Reference source not found. illustrates a cross-sectional view of a wafer-level chip-scale packaged (WLCSP) accelerometer, highlighting the vertical integration of the MEMS sensing structure and the signal-processing ASIC. It also shows the absence of leadframes and molded encapsulation, with electrical and mechanical connections provided through wafer-level bonding and solder interconnects.

WLCSP accelerometers offer extremely small footprints, often 2 $mm \times 2\ mm$ or smaller, along with improved electrical performance and lower mechanical noise coupling. These advantages make them well suited for wearables, earbuds, and compact IoT devices.

This illustration shows how the sensing structure and signal-processing ASIC are stacked within a chip-scale package, enabling a highly compact design while preserving separation between the mechanical sensing domain and the processing circuitry.

Beyond reducing package size, closer integration between the MEMS sensing structure and the signal-processing ASIC can improve electrical performance, reduce interconnect-related parasitics, and enhance device robustness. Shorter signal paths and tighter coordination between sensing and processing circuitry can also improve consistency across devices.

Further miniaturization will continue through advances in wafer-level stacking, tighter MEMS-ASIC integration, and more efficient die layouts. Likely developments include broader adoption of sub-2 mm packages, improved thermal sensing and compensation, reduced cross-axis interference, and expanded use of advanced interconnect technologies. These packaging advances will continue to expand where accelerometers can be used, especially in space-constrained products that demand low power, compact size, and reliable long-term performance.

7.5 Emerging Accelerometer Applications

As accelerometers become smaller, more accurate, more power efficient, and increasingly intelligent, they are finding their way into new application areas. These emerging use cases build directly on the same core capabilities discussed throughout this book, sensing acceleration, tilt, vibration, and motion patterns, but apply them in more specialized and demanding contexts. Rather than replacing existing applications, these trends expand the role accelerometers play in modern systems.

7.5.1 Advanced Wearable Health Monitoring

Next-generation wearable devices are using accelerometers for more than basic step counting or activity detection. Increasingly, they support posture and gait analysis, early detection of movement disorders, tremor characterization, breathing motion monitoring, and early indicators of fall risk.

Accelerometers are well suited to this domain because of their ultra-low power operation, long-term stability, ability to detect subtle motion, and compatibility with compact, body-worn form factors. These advances enable continuous monitoring for chronic disease management, rehabilitation, and preventative care.

Figure 7.10 illustrates an emerging class of wearable health devices that use ultra-low-power accelerometers to monitor posture, micro-movements, and breathing-related motion over extended periods. When placed on the upper body or other stable body locations, a small adhesive patch can detect tilt, subtle movement patterns, and long-term changes in motion behavior. These signals enable posture assessment, gait analysis, tremor characterization, and early indicators of fall risk or respiratory changes.

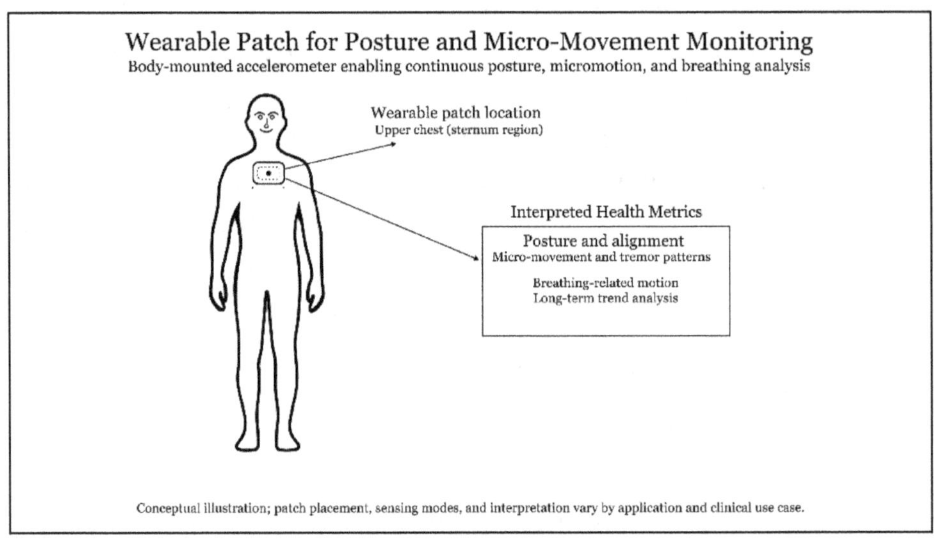

Figure 7.10 Wearable Patch for Posture and Micro-Movement Monitoring.

A chest-mounted accelerometer can monitor posture by tracking how the gravity vector projects onto the sensor axes as the upper trunk changes orientation. From these changes, the system can estimate upright posture, leaning forward, leaning back, and leaning to one side over time.

7.5.2 Trend Toward Edge-Based Industrial Condition Monitoring

A major emerging trend in industrial sensing is the shift from centralized vibration monitoring toward edge-based nodes that combine low-power accelerometers with local processing and wireless connectivity. Instead of transmitting raw vibration data continuously, these next-generation nodes classify motion locally and report only meaningful events, summaries, or anomalies.

Accelerometers in these systems support vibration anomaly detection, fault classification, estimation of remaining useful life, and adaptive thresholding based on machine operating cycles. This trend is already visible in smart factories and distributed industrial IoT deployments and is expected to expand as sensor intelligence increases.

7.5.3 Robotics and Autonomous Systems

Another important trend is the growing use of accelerometers in smaller, lower-cost autonomous and collaborative robotic platforms, where tighter integration and lower power operation are becoming increasingly important. Improvements in noise performance, drift behavior, timing stability, and synchronization with gyroscopes and vision systems enable smoother motion and more reliable control.

These benefits are especially important for small robots, drones, and service robots, where size, weight, and power constraints limit sensor options. Tighter IMU integration and low-power standby modes further expand deployment opportunities.

7.5.4 Smart Home and Appliance Intelligence

Smart-home sensing is also shifting toward more ambient, always-on interaction models in which low-power accelerometers provide background awareness with minimal user effort. These sensors can detect movement, orientation changes, vibration patterns, and everyday interaction cues, supporting maintenance diagnostics, safety monitoring, and new control methods such as knock or tilt gestures.

Future accelerometers will enhance these capabilities through better vibration classification, context-aware detection, and ultra-low-power always-on sensing that integrates seamlessly with home automation systems.

7.5.5 AR, VR, and Human–Machine Interfaces

As AR and VR devices become smaller, lighter, and more immersive, accelerometers are being pushed toward lower-latency and higher-fidelity motion sensing. Emerging use cases include micro-gesture detection, motion smoothing for hand tracking, synchronization with haptic (touch-based) feedback, and improved drift correction through tighter sensor fusion.

These applications demand stable orientation estimation and predictable timing, supported by tighter integration, improved calibration, and faster on-device processing in next-generation accelerometer systems.

7.5.6 Environmental and Structural Monitoring

A growing trend in infrastructure sensing is the use of distributed accelerometer networks for long-term structural and environmental monitoring. Accelerometers are increasingly being deployed in buildings, bridges, pipelines, and towers to track vibration, tilt, and gradual changes in structural behavior. Improvements in noise performance, long-term stability, and embedded intelligence are making it easier to detect subtle changes associated with wear, deformation, and environmental stress.

Another important trend is the shift from continuous raw-data collection toward low-power edge monitoring. Small sensor nodes can now operate for long periods using batteries or energy harvesting, while local processing reduces the need for constant data transmission and allows only summarized features, anomalies, or validated events to be reported upstream.

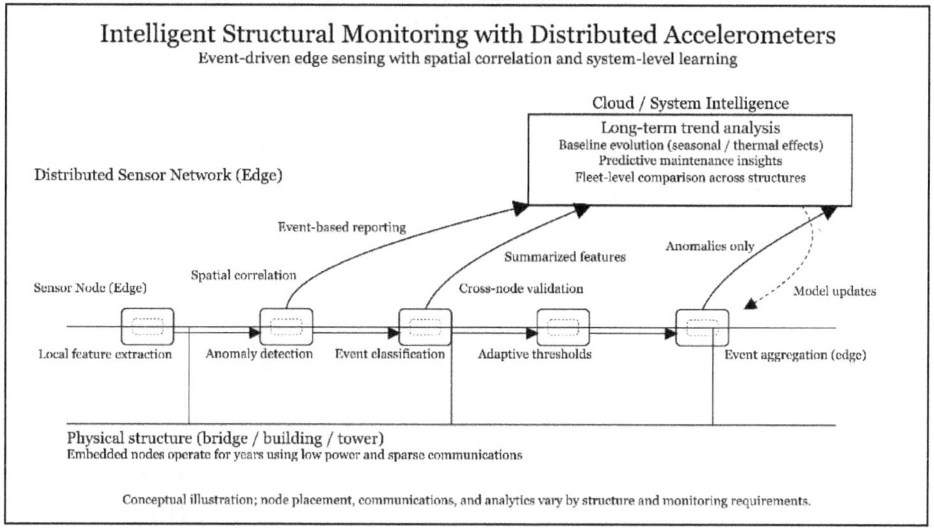

Figure 7.11 Intelligent Structural Monitoring Networks Using Distributed Accelerometers.

In **Figure 7.11**, each accelerometer sensor node represents the same edge device deployed at a different location along the structure. Rather than streaming raw acceleration continuously, nodes perform progressively richer local processing, including feature extraction, anomaly detection, event classification, and adaptive thresholding based on operating conditions.

Neighboring nodes exchange limited information to correlate measurements spatially and validate detected events, improving robustness and reducing false alarms. Some nodes additionally aggregate and refine event information before transmitting only summarized features or confirmed anomalies upward for system-level analysis. This distributed architecture allows large structures to be monitored efficiently over long time periods while minimizing power consumption, data transmission, and maintenance overhead.

All sensor nodes shown in **Figure 7.11** (five in this example) share a common and identical hardware design, including the accelerometer, local processing, and communication interfaces. The different labels associated with each node indicate logical edge-processing roles within the network, such as feature extraction, anomaly detection, or event aggregation, rather than distinct device types or specialized hardware.

Across these examples, accelerometer technology is advancing less through entirely new sensing physics than through better integration, lower power consumption, improved local processing, and more application-specific system design. As these trends continue, accelerometers will remain foundational motion sensors while expanding into a wider range of health, industrial, consumer, and infrastructure applications through more intelligent, efficient, and tightly integrated system roles.

7.6 Takeaways

The evolution of accelerometer technology is defined by steady, practical refinement rather than disruptive change. Advances in power efficiency, on-sensor intelligence, integration, and packaging build on established MEMS principles while expanding the range of systems that can benefit from motion sensing. These developments reflect the real constraints of modern products, including limited power budgets, compact form factors, and growing expectations for intelligent behavior.

Across the trends discussed in this chapter, a common engineering direction emerges. Accelerometers are increasingly designed as system-aware components rather than passive sensors.

Ultra-low-power operation enables continuous, always-on monitoring. Integrated event detection and embedded AI reduce system-level processing and energy consumption. On-sensor fusion simplifies motion tracking and improves timing, calibration, and reliability. Miniaturization and tighter MEMS–ASIC integration allow sensors to be embedded in places and products that were previously impractical.

At the same time, emerging applications demonstrate how these improvements translate into real value. Health monitoring moves beyond basic activity tracking toward continuous, clinically meaningful observation. Industrial systems shift from reactive maintenance to predictive and adaptive monitoring at the edge. Robotics, drones, and immersive interfaces benefit from more stable and responsive inertial sensing. Infrastructure and environmental monitoring extend motion sensing into long-term, low-power deployments.

Taken together, these trends reinforce an essential conclusion: accelerometers are not being displaced by newer technologies, but are becoming more capable, more efficient, and more tightly integrated into intelligent systems. Their role now extends beyond simple motion detection toward higher-level interpretation, autonomy, and context awareness.

With this perspective on current and emerging trends, the technical discussion of accelerometers reaches its conclusion. The remaining sections of this book include a Quick Reference Summary of key concepts, a Glossary of technical terms used throughout, and Suggested Reading & Resources for further exploration. Together, these sections are intended to help readers consolidate their understanding and apply accelerometer technology more confidently in real-world designs.

Quick Reference Summary

This Quick Reference Summary provides a fast, high-level review of the book's most important concepts. It is designed to help readers quickly recall key ideas, formulas, and best practices without revisiting entire chapters.

What Accelerometers Measure

Acceleration is the change in velocity over time. Gravity is a constant 1 g downward vector commonly used for tilt sensing. Vibration consists of periodic acceleration patterns across frequencies. Static measurements relate to orientation and tilt, while dynamic measurements capture motion, impact, and vibration. Accelerometers do not directly measure rotation; they measure only linear acceleration.

Core Sensing Principle (Capacitive MEMS Accelerometers)

A proof mass suspended by springs moves in response to acceleration. This movement changes the capacitance between interlocking comb fingers. The ASIC converts the capacitance change into a voltage and then into a digital output. Key relationships include the steady-state relationship between proof-mass displacement, applied acceleration, and spring stiffness, as well as the dependence of capacitance on electrode overlap area and the inverse dependence on electrode gap. Even nanometer-scale displacements generate measurable signals.

Important Specifications and What They Mean

Full-scale range defines the maximum measurable acceleration; in practice, the smallest range that avoids saturation should be selected. Sensitivity is expressed in $\frac{LSB}{g}$ or $\frac{mV}{g}$, with higher sensitivity providing finer resolution. Resolution is the smallest detectable change in acceleration and improves with lower full-scale range, higher ADC resolution, and lower noise.

Noise density, expressed in $\frac{\mu g}{\sqrt{Hz}}$, indicates how much noise is present for each square root of bandwidth, while RMS noise represents total noise across the selected bandwidth. Bandwidth defines the maximum frequency that can be measured accurately. ODR specifies how often new data is produced and must satisfy $ODR \geq 2 \times BW$ to meet the Nyquist condition.

Temperature drift affects offset and sensitivity across temperature, often requiring compensation. Cross-axis sensitivity occurs when motion on one axis appears on another and should be minimized for precise applications. Shock and vibration ratings define survivability rather than measurement capability.

Filtering Essentials

Low-pass filtering reduces noise, introduces delay, and removes high-frequency components, making it suitable for tilt, posture, and slow motion. High-pass filtering removes gravity and isolates dynamic motion but is not suitable for tilt estimation. Band-pass filtering is used for vibration analysis. The key principle is that filtering must match the motion characteristics of the application.

Power Modes and Low-Power Operation

Active mode provides full performance. Low-power mode reduces bandwidth and ODR, often at the cost of higher noise. Wake-on-motion mode enables sub-microamp detection using simplified logic. Standby mode minimizes current consumption. It is important to check startup settling time and configuration retention when switching modes.

Calibration Essentials

Offset calibration corrects the sensor's zero-g reading. Sensitivity calibration corrects scaling errors such as variation in $\frac{LSB}{g}$. Alignment calibration addresses axis misalignment. Temperature compensation corrects offset and sensitivity changes across temperature. Calibration should be performed only after the data has stabilized.

Common Mistakes to Avoid

Common errors include choosing an incorrect full-scale range, misunderstanding noise behavior, confusing ODR with bandwidth, misusing filtering, poor sensor placement that introduces stress, ignoring temperature effects, setting incorrect interrupt thresholds, overreliance on single raw samples instead of sequences, and neglecting timing and latency introduced by FIFOs and filters.

Practical Formulas

Displacement under acceleration is given by $x = (m \times a) / k$. RMS noise is $RMS = Noise\ Density \times \sqrt{BW}$. Acceleration magnitude is $mag = \sqrt{a_x{}^2 + a_y{}^2 + a_z{}^2}$. Tilt angles from acceleration can be computed as $pitch = arctan(\frac{a_x}{\sqrt{a_y{}^2 + a_z{}^2}})$ and $roll = arctan(\frac{a_y}{\sqrt{ax^2 + az^2}})$.

Yaw cannot be determined from accelerometer data alone because rotation about the gravity vector does not change the gravity projection on the sensor axes; it therefore requires an external heading reference such as a magnetometer or gyroscope-based estimation.

Application Matching Guidelines

Tilt and orientation sensing favor low bandwidth in the range of 2 Hz to 5 Hz, low noise, and stable offsets. Wearables prioritize low power, wake-on-motion capability, and moderate filtering. Drones and robotics require bandwidths in the hundreds of Hz, high ODR, and low latency. Industrial vibration monitoring often requires bandwidths from about 200 Hz into the kHz range, along with high dynamic range and careful anti-aliasing. Medical monitoring emphasizes low noise, long-term stability, and predictable drift behavior.

Sensor Fusion Basics

Accelerometers measure gravity and linear acceleration, gyroscopes measure rotation, and magnetometers provide heading information. Fusion algorithms such as Kalman and complementary filters combine these sensors to achieve stable orientation estimation, improved heading estimation, and reliable motion classification.

Future Trends to Watch

Key trends include ultra-low power MEMS, on-sensor machine learning, built-in fusion engines, smaller multi-sensor packages, smart context-aware detection, and expanded use in wearables, IoT, robotics, and infrastructure monitoring.

Glossary

−3 dB point: The frequency at which a system's output power falls to half of its reference value, corresponding to about 70.7% of the output amplitude. It is commonly used to define the practical cutoff or bandwidth limit of a sensor or filter.

Acceleration: The rate at which velocity changes over time. Measured in m/s^2 or g ($1\ g \approx 9.81\ m/s^2$).

Accelerometer: A sensor that measures acceleration along one or more axes, including dynamic motion, vibration, and the effect of gravity.

Active Mode: A sensor operating mode that provides wider bandwidth and ODR options, along with lower noise, but consumes more current.

ADC (Analog-to-Digital Converter): Circuit that converts analog voltage signals into digital values.

ADC Resolution: The number of discrete digital output levels produced by the analog-to-digital converter, often expressed in bits.

Aliasing: A signal-processing error that occurs when sampling frequency (ODR) is too low for the highest frequency in the signal, causing false low-frequency components.

Analog Front End (AFE): The analog circuitry that conditions the sensor signal before digitization, including amplification, filtering, and conversion of small physical signal changes into measurable electrical signals.

ASIC (Application-Specific Integrated Circuit): A custom integrated circuit designed for a specific function, integrating the analog front end, analog-to-digital conversion, signal processing, calibration, embedded features, and communication interfaces within a single device.

Augmented Reality (AR): A technology that overlays digital information, such as images, graphics, or text, onto the real-world environment in real time. AR systems combine sensor data, including motion and orientation measurements, with camera input to align virtual content with the physical world. In motion-sensing applications, accelerometers and gyroscopes help track device movement and maintain stable, accurate placement of virtual elements.

Automotive Grade 1: An $AEC\text{-}Q100$ qualification grade indicating an operating temperature range of $-40\ °C$ to $+125\ °C$, suitable for under-hood or high-temperature automotive environments.

Automotive Grade 2: An $AEC\text{-}Q100$ qualification grade indicating an operating temperature range of $-40\ °C$ to $+105\ °C$, suitable for general automotive environments outside high-temperature or under-hood conditions.

Band-Pass Filter (BPF): A filter that passes a selected range of frequencies while attenuating lower and higher frequency components, commonly used in vibration analysis.

Bandwidth (BW): The frequency range over which the accelerometer maintains a consistent response, typically referenced near the $-3\ dB$ point (where signal power drops to half, or amplitude to about 70.7% of its low-frequency value). It is shaped by the sensor design and commonly limited by analog and digital filters to balance signal fidelity and noise.

Bias / Offset: The accelerometer's output when no acceleration is applied ($0\ g$), excluding gravity.

Bias Drift / Offset Drift: Slow change in the zero-g output over time or temperature.

Brownian Noise: Fundamental mechanical noise due to microscopic thermal motion (random heat-driven movement) of the proof mass.

Calibration: A process used to measure and correct systematic errors in accelerometer output, such as offset, sensitivity, alignment, temperature drift, and in some cases non-linearity.

Capacitive MEMS: A microfabricated structure using moving capacitor plates to detect displacement caused by acceleration.

Centripetal Acceleration: Acceleration toward the center of a circular motion; sensed by accelerometers during rotation.

Comb Fingers: Interlocking capacitor plate structures inside a MEMS device that change capacitance when the proof mass moves.

Complementary Filter: A lightweight sensor fusion algorithm combining accelerometer and gyroscope data for stable orientation estimation.

Cross-Axis Sensitivity: The output of a sensor channel in response to acceleration applied perpendicular to its sensitive axis, typically expressed as a percentage of the applied acceleration.

Cutoff Frequency: The frequency at which a filter's output is reduced to the $-3\ dB$ point, marking the beginning of signal attenuation.

Data-Ready Interrupt (DRDY): A signal that indicates a new accelerometer sample is available.

Damping: Resistance to motion caused by gas or mechanical forces inside the MEMS cavity.

Dead Reckoning: The process of estimating current position from a known starting point by integrating measured motion over time. In practice, small sensor bias and noise errors accumulate through integration, so drift grows over time and external references are often needed for correction.

Decimation: The process of reducing the sampling rate of a signal by downsampling after applying a low-pass filter to remove high-frequency content and prevent aliasing.

Differential Capacitor: A pair of capacitors arranged so that one increases while the other decreases as the proof mass moves, producing a differential signal that improves noise rejection, reduces common-mode effects, and enhances measurement stability.

Digital Filtering: Processing applied to the digital output after the ADC to smooth the signal, reduce noise, or shape the bandwidth for the intended application.

Displacement: The movement of the proof mass relative to the sensor frame in response to acceleration, often measured in nanometers in MEMS accelerometers.

Duty Cycling: A power-saving technique in which sensor functions or circuitry are turned on and off periodically to reduce average power consumption while maintaining sufficient responsiveness.

Edge Computing: Processing data locally near the sensor rather than transmitting raw data to a remote processor or cloud system, reducing latency, bandwidth use, and power consumption.

Effective Resolution: The smallest acceleration change that can be measured reliably after accounting for noise, filtering, and other practical limitations.

Electrostatic Self-Test: A built-in test feature that applies an internal electrostatic force to the proof mass to verify the integrity of the mechanical structure and signal chain.

Event Detection: The identification of specific motion conditions, such as tap, free fall, tilt, or impact, based on predefined thresholds, timing logic, or pattern recognition.

FFT (Fast Fourier Transform): A mathematical method that converts time-domain accelerometer data into the frequency domain, allowing vibration content and dominant frequencies to be analyzed.

FIFO (First-In, First-Out buffer): Internal memory that stores accelerometer samples in the order they are acquired, helping reduce MCU workload, support burst reads, and preserve timing continuity.

Filter Latency: The delay introduced by analog or digital filtering because the filter requires time or multiple samples to respond fully to changes in the input signal.

Full-Scale Range (FSR): The maximum acceleration range the sensor can measure before the output saturates or clips.

Fusion Engine: Hardware, firmware, or embedded processing that combines accelerometer, gyroscope, and sometimes magnetometer data to estimate orientation, motion state, or other higher-level outputs.

g: Standard unit of acceleration based on earth's gravity, where $1\,g \approx 9.81\,m/s^2$.

Global Navigation Satellite System (GNSS): A general term for satellite-based navigation systems that provide global positioning, velocity, and timing information, including GPS, GLONASS, Galileo, and BeiDou.

Global Positioning System (GPS): A satellite-based navigation system developed by the United States that provides positioning, velocity, and time information to receivers on or near Earth.

Gravity Vector: The acceleration vector associated with gravity, whose projection onto the sensor axes is used to estimate tilt and orientation.

Gyroscope: A sensor that measures angular velocity, often used together with accelerometers in inertial measurement and sensor fusion systems.

High-Pass Filter (HPF): A filter that attenuates low-frequency components, such as gravity or slow drift, while allowing higher-frequency motion to pass.

Hysteresis: A difference in sensor output or threshold behavior depending on whether the input is increasing or decreasing.

I²C/SPI: Common digital communication interfaces used to configure accelerometers and read their output data.

I³C: A newer serial communication interface designed as a successor to I²C, offering higher data rates, lower power consumption, in-band interrupts, and dynamic device addressing while remaining backward compatible with I²C devices.

IMU (Inertial Measurement Unit): A device that combines an accelerometer, gyroscope, and sometimes a magnetometer to measure motion and support orientation estimation.

Interrupt: A signal from the sensor to the host processor indicating that a specified event or condition, such as motion, tap, free fall, or data ready, has occurred.

In-Run Bias Stability: The short-term stability of sensor offset during continuous operation under constant conditions.

Kalman Filter: A mathematical algorithm that combines multiple sensor inputs and their uncertainties to estimate state variables such as orientation, velocity, or position.

Latched Interrupt: An interrupt output that remains asserted after an event occurs until it is explicitly cleared by the host processor, helping ensure short-duration events are not missed.

Latency: The delay between a physical motion or event and the corresponding response, output, or recognition by the sensor or system.

Least Significant Bit (LSB): The smallest digital step in an output value, representing the smallest increment that can be expressed by the sensor's digital output.

Linear Acceleration: Acceleration excluding gravity, often estimated by removing the gravity component from the measured accelerometer output.

Low-Pass Filter (LPF): A filter that attenuates high-frequency components, reducing noise while preserving lower-frequency motion.

Low-Power Mode: A sensor operating mode that reduces power consumption, typically by lowering bandwidth, ODR, or internal processing activity.

Magnitude: The magnitude of acceleration across all measured axes, commonly calculated as $\sqrt{a_x^2 + a_y^2 + a_z^2}$.

Mechanical Resonance: The natural frequency at which the proof mass tends to oscillate, setting an upper limit on usable bandwidth and influencing dynamic response.

MEMS (Micro-Electro-Mechanical Systems): Microscale mechanical structures and electrical components fabricated using semiconductor manufacturing processes.

Misalignment: Deviation of the sensing axes from perfect orthogonality, which can introduce measurement error between axes.

Motion Classification: The process of using accelerometer data to identify specific movement types or motion states, such as steps, gestures, or activity patterns.

Noise Density: A measure of how much random noise a sensor produces per unit bandwidth, indicating how noise increases as the measurement bandwidth widens. It is typically expressed in $\frac{\mu g}{\sqrt{Hz}}$ or $\frac{mg}{\sqrt{Hz}}$ and is used to estimate total RMS noise over a given frequency range.

Nyquist Condition: The requirement that the sampling rate be at least twice the highest frequency of interest in the signal to avoid aliasing.

ODR (Output Data Rate): The rate at which the accelerometer produces new output samples, usually expressed in Hz.

Open Drain: An output configuration in which the device can pull the signal line low but cannot drive it high; an external pull-up resistor is required to define the high logic level and allow multiple devices to share the line safely.

Orientation: The direction or angular position of the device relative to gravity or another reference frame.

Orientation Estimation: The process of determining the device's angular position relative to gravity or a reference frame, often using sensor fusion.

Pitch: The rotation of a device about its lateral axis (side-to-side axis), typically corresponding to forward or backward tilt. In accelerometer-based systems, pitch is estimated from the projection of the gravity vector onto the sensor axes.

Power-Down Mode: The lowest-power sensor mode in which measurement activity is disabled and output data is unavailable.

Power-Up Time: The interval from initial power application until the device reaches its defined operating state, including supply stabilization, voltage ramp-up, and internal reset.

Proof Mass: The movable mass inside a MEMS accelerometer that responds to acceleration and produces the displacement used for sensing.

Pull-Up Resistor: A resistor connected between a signal line and a supply voltage to ensure the line defaults to a defined logic high level when not actively driven, commonly used with open-drain or open-collector outputs.

Quantization Noise: Error introduced by representing continuous signals with discrete digital values, typically appearing as a small uncertainty in the output.

Response Time: The time it takes for the accelerometer output to begin changing after an acceleration is applied, reflecting how quickly the sensor reacts to input changes.

RMS Noise: The total effective random noise at the sensor output within the selected bandwidth, expressed as a root-mean-square value. It reflects the combined noise level seen in the measurement and is typically estimated from noise density over the chosen bandwidth.

Roll: The rotation of a device about its longitudinal axis (front-to-back axis), typically corresponding to left or right tilt. In accelerometer-based systems, roll is estimated from the projection of the gravity vector onto the sensor axes.

Sample-and-Hold: An internal timing technique in which the sensor signal is sampled and held temporarily stable during analog-to-digital conversion.

Sensitivity (Scale Factor): The output change produced per unit of acceleration, such as LSB/g, mg/LSB, or mV/g.

Sensitivity Drift: The change in sensor sensitivity with temperature or operating conditions, causing the scale factor to vary over time or across the temperature range.

Sensor Fusion: The process of combining data from multiple sensors to improve accuracy, stability, or robustness beyond what a single sensor can provide.

Settling Time: The time required for the accelerometer output to reach and remain within a specified tolerance band around its final value after a transient such as overshoot or ringing has decayed.

Shock Rating: The maximum acceleration the sensor can survive without permanent damage, usually specified as a peak shock level over a defined duration.

Standby Mode: A low-power operating mode in which most internal circuitry is disabled and active measurement is suspended.

Startup Time: The time required after power-up or after enabling a measurement mode for the accelerometer output to become valid and stable for use, including internal initialization and settling.

Tap/Double-Tap Detection: Built-in event-detection functions that identify short impulses or repeated tap patterns based on programmable thresholds and timing logic.

Temperature Compensation: Adjustment of the sensor output to reduce errors caused by temperature-dependent changes such as offset drift or sensitivity drift.

Thermal Noise: Random noise caused by thermal energy in mechanical or electrical sensor components.

Tilt: Orientation relative to gravity, commonly expressed as an angle derived from accelerometer measurements.

Time Window: A sliding or fixed group of consecutive samples used for signal analysis, averaging, or pattern recognition.

TinyML: Machine-learning models optimized to run on low-power microcontrollers or embedded sensor ASICs with limited memory and processing resources.

Vector Magnitude: See **Magnitude**, the combined acceleration across all measured axes.

Virtual Reality (VR): A technology that immerses users in a fully computer-generated environment, often using accelerometers and gyroscopes for motion tracking, orientation sensing, and user interaction.

Wake-on-Motion: A low-power mode in which the accelerometer monitors motion and wakes the host processor only when motion exceeds a defined threshold.

WLCSP (Wafer-Level Chip-Scale Packaging): A packaging approach in which the MEMS sensing structure and signal-processing ASIC are integrated and packaged at the wafer level, enabling very small and compact sensor packages.

Yaw: Rotation about the vertical axis of a device or body. Unlike pitch and roll, yaw cannot be determined from accelerometer data alone because rotation about the gravity vector does not change the gravity projection on the sensor axes.

Suggested Reading & Resources

This section highlights some of the authoritative resources that provide deeper insight into accelerometers, MEMS design, signal processing, and sensor fusion. These materials complement the concepts introduced in this book and serve as long-term references for students, engineers, and developers who wish to expand their understanding of modern inertial sensing technology.

Manufacturer Publications

Sensor manufacturers publish application notes, design guides, and technical references that provide practical guidance on configuration, calibration, filtering, power optimization, and system integration. These documents are often written by the engineers who design the sensors and are among the most reliable sources of real-world implementation guidance.

Readers are encouraged to explore application notes and design resources available on sensor manufacturers' official websites or through technical documentation portals. These materials are regularly updated to reflect new devices, evolving best practices, and common implementation challenges.

Modern accelerometer datasheets are also valuable engineering references. Beyond basic specifications, they often include detailed information on noise behavior, power modes, filtering options, timing, electrical interfaces, and recommended operating conditions. Reviewing datasheets from multiple manufacturers helps engineers compare design approaches, understand practical trade-offs between performance, power consumption, robustness, and cost, and ensure that specifications and application guidance remain current.

Professional and Academic Resources

In addition to manufacturer documentation, books and publications authored by recognized experts in MEMS technology, microsystems design, and inertial sensing remain valuable long-term references. Professional organizations and academic publishers also provide educational material that reflects established theory and industry practice.

MEMS / Fabrication and Sensors

Stephen Senturia - Microsystem Design
A foundational work on MEMS devices, covering mechanical structures, sensing principles, and system-level design.

Marc Madou - Fundamentals of Microfabrication and Nanotechnology
A detailed reference work on microfabrication processes used to create MEMS and microscale sensing structures.

Gregory Kovacs - Micromachined Transducers Sourcebook
An accessible yet thorough introduction to MEMS transducers, including accelerometers, gyroscopes, and resonant devices.

Signal Processing & Measurement Techniques

Richard Lyons - Understanding Digital Signal Processing
A practical guide to DSP concepts, with clear explanations applicable to filtering and processing accelerometer signals.

Dan Simon - Optimal State Estimation
Covers Kalman filtering and related estimation algorithms widely used in control systems, navigation, and IMU sensor fusion.

IMU, Sensor Fusion, and Inertial Navigation

Sebastian Madgwick - An efficient orientation filter for inertial and inertial/magnetic sensor arrays
Widely used introductory work on low-computation orientation filters.

Paul D. Groves - Principles of GNSS, Inertial, and Multi-sensor Integrated Navigation Systems
A comprehensive reference on inertial navigation and sensor fusion, including Kalman filtering and real-world IMU system integration.

How to Use These Resources

These references are not required to understand the content of this book. They are intended for readers who want to study MEMS structures in greater detail, design custom filtering or signal-processing algorithms, build advanced IMU systems, explore industrial or medical applications more deeply, or better understand the engineering behind commercial accelerometers. Together, these resources provide a balanced foundation for further study in inertial sensing, accelerometer design, embedded systems, and real-world application development.